T0213761

Additive Manufacturing Solutions

Sanjay Kumar

Additive Manufacturing Solutions

 Springer

Sanjay Kumar
Gumla, Jharkhand, India

ISBN 978-3-030-80785-6   ISBN 978-3-030-80783-2   (eBook)
https://doi.org/10.1007/978-3-030-80783-2

This Springer imprint is published by the registered company Springer Nature Switzerland AG
The registered company address is: Gewerbestrasse 11, 6330 Cham, Switzerland

Contents

Abbreviations

AM	Additive manufacturing
BP	Bed process
BS	Bed system
CM	Conventional manufacturing
CSAM	Cold spray additive manufacturing
DDM	Direct digital manufacturing
DM	Digital manufacturing
DMD	Digital micromirror device
DP	Deposition process
DS	Deposition system
EPBF	Electron beam powder bed fusion
IC	Investment casting
IM	Injection molding
LPBF	Laser powder bed fusion
PBF	Powder bed fusion
PBS	Powder bed system
PPBP	Photopolymer bed process
PPBS	Photopolymer bed system
RM	Rapid manufacturing
SS	Support structure
WAAM	Wire arc additive manufacturing

About the Author

Sanjay Kumar (Ph.D., KUL, Belgium) who worked at Utah State University (USA), National Laser Centre (South Africa), Shanghai Jiao Tong University (China), and York University (Canada), is the author of "Additive Manufacturing Processes", 2020, Springer, Cham. Email: skumarsdr@gmail.com

Chapter 1
Synonym

1 What Is Solid Freeform Fabrication?

AM is called solid freeform fabrication (SFF) because it can fabricate a solid object having form (shape) that is free from any limitation. Other manufacturing processes offer limited shapes, and are therefore not called solid freeform fabrication.

For example, in injection molding (IM), the shape of the cavity of mold restricts the form of a solid object fabricated; the object therefore has no freedom to take any form other than that given by the mold.

1.1 Disadvantage of Mold Based Process

Complexity in IM depends upon how complex a mold can be formed, how well the mold can be filled up during injection, and how well the mold allows the injected material to be ejected [1]. If there is incompetency at any stage of the process, it will affect the maximum complexity of the solid form that can be achieved.

Mold has disadvantages—if a thin wall or a small pin needs to be made, a mold having thin cavity is required that is not be possible to be fabricated with ease. Similarly, if a narrow hole needs to be made, it may not be possible to make a mold having narrow protrusion for such requirement. Even in that type of molds, it is not possible to make objects easily. Even if it is possible to make objects from such molds, the molds will not last long to justify cost investment. Thus, if there is a design which consists of such features, IM is not a right choice to make such objects.

1.2 Advantage of Mold Based Process

If the shape of an object in IM was not limited by the shape of the mold, it could have freedom to make any shape, but the freedom would not come free. The confining space of the mold has advantage that it does not let material go astray. In the absence of the confining space, right parameters need to be searched for the material to remain confined within a designated space. But due to the confining space offered by the mold, materials can be stacked rather fast. Otherwise, every bit of the material requires to be checked whether the material is at the right place before solidification or whether it is not shifted due to shrinkage after the solidification. When every bit of material takes attention, the process will not be as fast as in the presence of a confining space where such attention is not required. This is one of the reasons for the slowness of AM where every bit of material addition needs attention.

If there is something (e.g., mold) which gives a confining space, then after the fabrication of one object, another object can be fabricated and so on—this can go fast. This advantage is absent when there is no mold (or no confining space as an outcome). In AM, each object needs to be grown, even if the object to be fabricated is always of the same shape—there is no certainty that the two objects fabricated will have no difference due to the growth. Thus, a mold gives an advantage of repeatability. If the same object always needs to be fabricated, IM is a preferred choice [2]. This is the advantage gained with mold based process.

Thus, mold has given advantages. It has made IM faster than AM. It has given repeatability [3], which has caused IM rather than AM as a suitable process for mass production.

But, if a customized product rather than mass production is priority, if fabrication of a product is more important than the speed of the fabrication, if an old product having guarantee in property due to the confining space is less preferred than an innovative product, AM is better than IM.

2 What Does Rapid of Rapid Manufacturing Imply?

If a digital file needs to be changed frequently to make new products or to verify the design of a product, it can be expensive and time-consuming, if done through injection molding (IM). Since with each change in file, a new mold needs to be fabricated first for letting the fabrication of a new product to follow.

There is no requirement of a mold (an intermediate step) in AM. The lack of requirement allows AM to reach physical object stage from digital file stage without going through the intermediate step. Thus there is one step less in AM in comparison to IM, which is mold-based conventional manufacturing (CM). The requirement of fewer steps makes AM a fast manufacturing process. That is why AM is called rapid manufacturing (RM), but it does not imply that AM in general is always faster than CM. It only implies that AM is rapid manufacturing because the delay associated with the intermediate step does not exist in AM. Therefore, AM can be faster than those CM which get delayed on going through intermediate steps.

AM is RM because it can rapidly fabricate physical objects by changing digital files (design) which is not possible in IM. But AM, for example, cannot be rapid manufacturing for making a big object starting from a small feedstock (for example, powder) when the same object can be made rapidly by other manufacturing processes such as forging (a CM). In this case, the absence of the intermediate step in AM does not help AM enough to be faster than CM.

In another example, if a complex engine part is planned to be manufactured through investment casting (IC), the part cannot be manufactured by a single step but requires additional steps of making casting patterns (refer Chap. 6). While in AM, the same part can be manufactured in a single step. Thus, AM is faster than IC since AM unlike IC does not need to have a casting pattern first. If the part is not small but big such as aircraft wing, AM may not be faster. Thus, AM is not always faster even if additional steps are not needed in AM while CM is not always slower even if it has to go through additional steps.

When the purpose of RM is to create a tool instead of any products, RM is called rapid tooling [4].

3   What Is Rapid Prototyping?

Rapid prototyping is one of the oldest and current applications of AM [5] (refer Chap. 6). AM is rapid prototyping when AM is used to fastly verify the appropriateness of a design and therefore to make a physical object that is not meant to be utilized as an end-use product but to serve as a medium to demonstrate the 3D form of the design.

3.1   Visual Prototype

When a physical object that will be used as a prototype needs to be fabricated, the object does not need to have as strong property as it will have when it will not be meant to be used as a prototype. If the prototype is a visual prototype, it must look like closer to the digital image but it does not need to have as high mechanical properties as that of a final object. The mechanical property of a visual prototype unlike that of a final object is considered high enough when the prototype is not prone to disintegrate on checking its appearance.

3.2   *Functional Prototype*

If a prototype is a functional prototype [6], it must have enough property to sustain the testing of functions, but it does not need to have as strong property as that of a final object that the functional prototype may eventually lead to. For example, when prototype of a nozzle is to be fabricated to test the flow of air through it, the prototype needs to have enough properties so that its functionality can be checked. These properties can be (1) enough strength so that it must not break due to the flow of air when it is fitted with an air source, (2) its inner surface must be as smooth as that of a final object to test the flow well. After checking the functionality for a required number of times, if the prototype breaks down, the prototype is not a right product to be used as a nozzle but is nevertheless a right prototype because it serves the purpose.

3.3   *Functional Prototype to Check All Properties*

If a functional prototype is fabricated to check mechanical properties of an object, the fabrication of the prototype will require as much time as that of a final object to match their properties. If the final object is of complex shape and the aim is only to check mechanical properties, simpler shape of the prototype (or test specimen shape) instead can work. In this case, fabrication of a prototype will require less time than that of the final object. This type of functional prototype is known as mechanical benchmark [7].

If the aim of the prototype is to check everything what a final object can furnish such as design, complexity, functionalities, and properties, then there will not be any difference between a prototype and a final object. The fabrication time in both cases will be the same.

Thus fabrication of all types of prototypes is easier than that of final objects except when functional prototype is fabricated to test all functionalities and properties. Since prototype-fabrication was easier, it was the earliest AM application when AM was not developed to fulfill the critical requirement of high value products but just to fulfill the lower requirement enough for a prototype.

4   What Is the Difference Between Additive Manufacturing (AM) and 3D Printing?

3D printing is another name for AM, and both names are used interchangeably. However, in practice, 3D printing is used differently.

The term 3D printing is used mostly in the following types of cases:

- When application of AM is more important than knowing technical details of AM. For example, in a primary or secondary school where it is more important to convey how various important pieces of items can be produced from a small adget than what are the various methods of manufacturing. If the term additive manufacturing is used, it may not be possible to convey the idea because students may not be aware of any types of manufacturing. But if 3D printing is used, conveying will be easier as students may be aware of both the words: 3D and printing.
- In non-technical environments where the term 2D printing is known more than the term manufacturing. In those cases, 3D printing can be intuitively understood as the next higher level of 2D printing. If the term additive manufacturing is used, it will require explanation why additive manufacturing is not the manufacturing of additives.
- In those technical environments where technical field is other than the manufacturing such as medical field or food industry where the term 3D printing is understood by majority of practitioners.
- For low value AM products when quality is not critical. For example, 3D printing is used in gaming industry since the quality of a model in the gaming industry is not as important as the quality of a part in aerospace industry.

5   What Are the Broad Categories of AM?

AM or additive layer manufacturing is divided into two broad categories: material bed process and material deposition process [8]. Material bed process and material deposition process are synonym for bed process (BP) and deposition process (DP), respectively. BP is synonym for bed based AM process while DP is synonym for deposition based AM process.

BP has categories such as powder BP, slurry BP, photopolymer bed process (PPBP), etc. Powder BP has categories such as metal powder BP, polymcr powder BP, etc. DP has categories such as solid DP, liquid DP (refer Chap. 7), wire DP, filament DP, powder DP, laser solid DP, laser powder DP, laser wire DP, electron beam wire DP (refer Chap. 5), and so on.

Systems for BP and DP are bed system (BS) and deposition system (DS), respectively. Thus BS is synonym for bed based AM system while DS is synonym for deposition based AM system. BS related with powder is powder bed system (PBS) that has categories such as metal PBS, polymer PBS, etc. BS related with photopolymer is photopolymer bed system (PPBS) (refer Chap. 8).

References

1. Masato, D., Sorgato, M., & Lucchetta, G. (2021). A new approach to the evaluation of ejection friction in micro injection molding. *Journal of Manufacturing Processes, 62,* 28–36.
2. Ituarte, I. F., Coatanea, E., Salmi, M., et al. (2015). Additive manufacturing in production: A study case applying technical requirements. *Physics Procedia, 78,* 357–366.
3. Wesemann, C., Spies, B. C., Schaefer, D., et al. (2021). Accuracy and its impact on fit of injection molded, milled and additively manufactured occlusal splints. *Journal of the Mechanical Behavior of Biomedical Materials, 114,* 104179.
4. Dimov, S. S., Pham, D. T., Lacan, F., & Dotchev, K. D. (2001). Rapid tooling applications of the selective laser sintering process. *Assembly Automation, 21*(4), 296–302.
5. Gebhardt, A. (2011). *Understanding additive manufacturing.* Ohio: Hanser publications.
6. Nieto, D. M., López, V. C., & Molina, S. I. (2018). Large-format polymeric pellet-based additive manufacturing for the naval industry. *Additive Manufacturing, 23,* 79–85.
7. Rebaioli, L., & Fassi, I. (2017). A review on benchmark artifacts for evaluating the geometrical performance of additive manufacturing processes. *International Journal of Advanced Manufacturing Technology, 93,* 2571–2598.
8. Kumar, S. (2020). *Additive manufacturing processes.* Cham: Springer.

Chapter 2
Advantage

1 What Are the Advantages of AM?

All advantages of AM are derived from the main advantage, i. e., freedom to add.

AM gives freedom to add material. This is the only way a part is fabricated in AM. Though the freedom is not blind—the freedom is not available to add materials at random but to add progressively in the direction of build, the freedom is again restricted to add in a plane before moving to add in the next plane in layerwise manufacturing, etc. The limited freedom that is left after these restrictions is still an advantage. Conventional manufacturing (CM) such as sintering, molding, casting, etc. which add materials does not have such freedom and such advantage as a consequence.

The freedom to add material gives rise to creativity (refer Chap. 7). Material can be added anywhere on a plane in whatever amount possible giving rise to various geometries. If the material is not added, it is because the material is not fit enough or the process is not developed for that material. But addition of any type of materials (and giving rise to any geometries) is not in conflict with the principle of AM.

If the variety of materials is changeable during the course of addition, the addition will give rise to a multi-material part having many properties, or one unique property [1], or a functionally graded part, or a composite part (refer Chap. 5) made up of different materials at different locations (of the part) having potential to serve different functions.

1.1 Types of Advantages

AM gives the following three advantages (Fig. 2.1):

1. Freedom to design

© The Author(s), under exclusive license to Springer Nature Switzerland AG 2022
S. Kumar, *Additive Manufacturing Solutions*,
https://doi.org/10.1007/978-3-030-80783-2_2

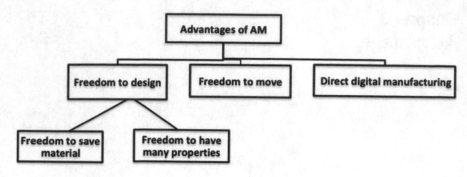

Fig. 2.1 Advantages of AM

2. Freedom to move
3. Direct digital manufacturing (DDM) (refer Chap. 9)

The first advantage (freedom to design) gives rise to two advantages: freedom to save material and freedom to have many properties.

1.2   Decoding Various Types

Freedom to Design and Freedom to Move

The advantage freedom to design comes from the absence of design-specific tooling while the advantage freedom to move comes from the compactedness of the system. If the system is not able to make design, it will lose its advantage of being moved and located. But this does not make advantage freedom to move to be dependent on the advantage freedom to design. For example, if an AM system is big and has more design capability than a small AM system, the small system has more freedom to move than the big system has.

Vice versa, freedom to design does not depend on the size of the system, it is thus independent of the advantage freedom to move.

Freedom to Design and Direct Digital Manufacturing

The advantage DDM depends upon the ability of AM system to be automated. Thus, if an AM system has freedom to design, it does not mean it can be automated as well. For example, both laser powder bed fusion (LPBF) and electron beam powder bed fusion (EPBF) have freedom to design but both have different problems to be automated. In LPBF, scan speed does not change laser power while in EPBF scan speed changes electron beam power. Thus automation in EPBF faces different difficulties than in LPBF. Hence, having same freedom to design does not imply that having same ease for automation.

Vice versa is also true, for example, both LPBF and laser powder deposition process (DP) can be automated well in spite of having different design capabilities. Thus having ability to be automated does not imply having capability to design.

While freedom to design depends upon what happens during building of a product, DDM requires not only what happens during building but what happens before and after building—how material will move to the processing zone and how product will leave the processing zone so that product fabrication will be continued. Thus freedom to design and DDM are two different aspects of AM. These different aspects are different advantages when seen from the perspective of CM.

Freedom to Design and Sub-Advantages

Since design is realized by adding material selectively, and material is saved because it is added selectively, freedom to save material comes from freedom to design. Since freedom to design allows to add material by choosing different materials which can lead to multi-material or multi-property product, the advantage freedom to have many properties comes from freedom to design.

Freedom to save and freedom to have many properties are different advantages while the latter advantage does not come from the former advantage. Freedom to save means to add material frugally which will result in a part having thinner sections and missing (unwanted due to improvement in design) sections. While freedom to have many properties means adding one material for one section and adding other material for other sections—it will not lead to minimal use of material but will give rise to any use of material to give many properties. Thus, freedom to have many properties will not necessarily lead to freedom to save material and vice versa. Hence, these two advantages are independent advantages but emanates from the advantage freedom to design and are thus its sub-advantages.

In the absence of freedom to design, these two advantages do not exist but it does not mean freedom to design will necessarily lead to freedom to save. It only means if there is freedom to save, it comes only from freedom to design.

2 How Does AM Give Freedom to Design?

The freedom to design is a relative term showing the freedom that did not exist earlier in CM is henceforth available in AM. For example, if a hollow cube having lattice surfaces (hollow lattice cube) (Fig. 2.2a) cannot be fabricated in machining but can be fabricated in powder bed fusion (PBF); there was no freedom to design it earlier (using machining) but with the advent of AM (using PBF) the freedom is henceforth available.

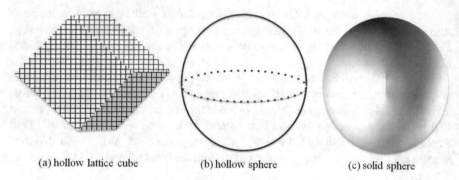

(a) hollow lattice cube   (b) hollow sphere   (c) solid sphere

Fig. 2.2 Schematic diagram of (**a**) hollow cube having lattice surface, (**b**) hollow sphere, and (**c**) solid sphere

2.1 AM Is Not Always Better

If a hollow sphere (Fig. 2.2b) needs to be made, the sphere though simpler than a hollow lattice cube is not possible to be manufactured in PBF [2]. As the closed sphere will not allow powder to be drained out. The hollow sphere can better be manufactured in other CM, i.e., metal forming. Thus for fabricating a hollow metallic sphere, CM is better than AM.

If a solid sphere (Fig. 2.2c) needs to be made, the sphere though again simpler in design than a hollow lattice cube is not economic to be manufactured in PBF. The solid sphere can better be manufactured in machining or sintering. Thus, for fabricating the solid metallic sphere, CM is again better than AM [3].

The advantage freedom to design is not applicable in all cases. If some design is not working in CM since the design is complex for a particular CM while the same complex design works in AM, it is the advantage of AM over CM. Therefore, the advantage of freedom to design is a relative advantage.

2.2 AM Is Better Because CM Fails

It is a relative advantage since it is seen from the perspective of CM, and it needs to be known beforehand whether the design failed in CM or not. In the case of a solid sphere or a hollow cube, the design did not fail in CM.

When the design did not fail in CM while the same design failed in AM, there was no advantage (of freedom to design) of AM over CM. When the design did not fail in CM while the same design did not fail again in AM, there was again no advantage (of freedom to design) of AM over CM.

To know the advantage, it is necessary that the design must fail in CM. Besides, it is again necessary that the design must fail in CM before even the design is tried in AM. It does not mean CM must not be improved or the aim should be to have a design necessarily fail in CM. It only means the design advantage offered by AM is not an advantage if AM is not developed enough to succeed in that design in which CM has already failed.

It does not mean the design advantage offered by AM does not have an independent value, or designing for AM necessarily requires pros and cons of designing in CM. It only means the design advantage of AM can only be known by comparing AM with CM.

Freedom to design, nevertheless, has a relative connotation—the design that was restricted earlier due to restriction in CM is no longer restricted because with the advent of AM that type of restriction is gone.

2.3 Restriction in CM

To realize the restriction in machining, a rectangular hollow space having two lattice boundary walls and one lattice internal wall needs to be fabricated (Fig. 2.3c). This can be made by hollowing the rectangular solid (Fig. 2.3a) in two sections

Fig. 2.3 Schematic diagram of (**a**) rectangular solid, (**b**) hollow sections of the solid, (**c**) hollow rectangular solid having three lattice walls

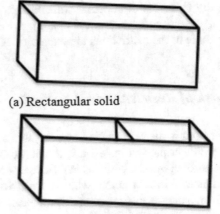

(a) Rectangular solid

(b) Two hollow sections of the solid

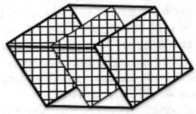

(c) Hollow rectangular solid having two side walls and one internal wall as lattice, side wall are not shown for clarity

(Fig. 2.3b). Hollowing the rectangular solid to half gives rise to four boundary walls of the hollow section and one internal wall. Two of the boundary walls and internal wall need to be made of lattice structures. In CM, for making lattice structure, there exists other suitable methods such as foaming, investment casting (IC), coating [4] but machining is selected instead to show its limitation.

Hollowing the solid by removing the material as desired by milling machine is the easiest part to make the desired structure. Making lattice structure on walls will require developing the tool path. If the lattice structure has unit lattice cell of variable densities, one milling tool will not be sufficient. If the cell has sharp corner, milling tool may not help. For making sharp corner of the cell, fine finish is required using non-conventional machining, for example, laser cutting. If all four walls have different types of lattices, both tool path programming and machining will take considerable time.

If struts of the some of the unit cells are thin, they may break by cutting force while they are made. The main problem will appear when attempt is made to create lattice structure in the internal wall. The tool will have no space to access the wall. If the rectangle is very big and machine is robust, the internal wall can be accessed but it will require changing the orientation of the rectangular and fixing it again for machining. If the rectangular is small, it is impossible to mill the internal wall and complete the fabrication of the part.

Difficulty to access the internal wall, fabricating sharp lattice, fabricating thin lattice due to cutting force, process planning with many tools, tool path programming are some of the restrictions imposed by CM.

2.4 Lack of Restriction in AM

If the same part is made in PBF (AM), the above restrictions that come along with CM are no longer present in AM and the part can be fabricated. This is the design advantage offered by AM which is available in AM but not in CM.

The advantage in AM is that what is a restriction in CM is not a restriction in AM. With an increase and variation in lattice cell density, there is an increase in tool path programming in CM. But in AM, there is no such corresponding increase in parameter optimization as long as the change in the lattice cell does not reach the limit of the process—this is an advantage in AM which allows to work with various lattice structures [5]. Non-reaching of tool to workpiece for fabricating a particular feature is a restriction in CM but not in AM. There is no such possibility in AM since after adding material, next addition happens in a building direction which has free space. If the next addition was in a reverse building direction, which does not happen in AM, the problem could creep in due to the lack of free space.

Managing a number of tools in machining or process planning with those tools or limitation posed by the cutting force due to those tools is the problem in CM but in absence of those tools in AM there are no such problems in AM (refer Chap. 5).

2.5 Comparison

It does not mean if AM does not have restrictions what CM has, AM does not have other restrictions as well. It also does not mean whatever restrictions are in AM are also the restrictions in CM. It only means if both restrictions that exist in AM and in CM are compared, it is found that from the design point of view CM has more restrictions than AM. Due to this comparison, there is more design to freedom in AM than in CM.

In PBF, if a particular aluminum alloy is found to work for fabricating a lattice structure, it will not be possible to make even a simple part by changing the composition of the aluminum alloy. Since in PBF, each alloy of a particular composition needs to be investigated, and if possible, optimized for manufacturability. It is a restriction in AM. But by taking aluminum alloy of different composition, machinability will not change in such a way that lattice structure will not be formed with new aluminum alloy in CM if it was formed earlier. Thus with CM, a range of composition of the same alloy has manufacturability. Changing the composition within a small range is not a restriction in CM but in AM. Consequently, it is not necessary that whatever is a restriction in AM will also be a restriction in CM.

But this advantage of CM over AM has little influence on design. For making such part having lattice structures, this advantage of CM will not lead to any ease in difficulty in tool path programming or access of tool to inaccessible sections of the workpiece or process planning with tools. The advantage of manufacturability of materials of CM over AM does not help CM overcome those restrictions for which CM is not better than AM.

In CM, for fabricating a lattice structure, circular rather than rectangular lattice cells can be easily fabricated since milling tool will have difficulty for making the corner of the rectangular cell. In AM, there is no such preference because of the absence of the milling tool. But in PBF, the size of lattice cell will be restricted though there will be no difference in making different shapes. The smallest size of the lattice cell will be decided by the type of beam spot size and resolution of the process, if the beam spot size is bigger, it will be difficult to control melting of powders to a small area.

Therefore, CM has restriction when shape of a cell changes from circular to rectangular but AM does not have such restriction. While AM has restriction when the size of a cell changes from bigger to smaller but CM does not have restriction as far as a smaller milling tool is available.

2.6 Design Advantage Is Not Enough

In PBF, for fabricating a lattice structure in a part, it is preferable that the lattice structure exists in the horizontal section of the part. The part needs to be built in such orientation that the section having lattice structure is horizontal to the build direction. If the section is in vertical direction rather than in horizontal, the lattice structure will have chance to bend or have higher surface roughness or crumble depending on the size. But, if the part has many sections having lattice structures, whatever be the build direction there will always be some sections which will not be in the horizontal direction. This may result in lattice structure of horizontal sections having different properties than the lattice structure in vertical direction. For those sections, which are neither in horizontal nor in vertical direction, the properties will again be different. Hence, with a change in orientation, there will be different toughness and propensity to crack [6]. These will bring questions on the reliability of the fabrication process and the repeatability of properties within a part.

These are the restrictions of AM but not of CM. In CM, lattice structure of different sections does not have such different properties due to the process, i.e., machining (if CM becomes capable enough to allow first those lattice structures or sections or parts to come up). IC is used to make lattice structures [7] but IC is a mold based CM while the comparison is with machining, a non-mold based CM.

Therefore these restrictions in AM do not belittle AM and take the design advantage from AM, these restrictions only imply that the design advantage of AM does not enable AM to have reliability (or repeatability or property) advantage.

2.7 Restriction Is an Advantage

To overcome the problem, PBF does not have option to stop the ongoing fabrication and change the build direction so that the horizontal section will become vertical and vice versa. This is the restriction of AM.

But stopping the ongoing fabrication and changing the workpiece orientation are not a restriction in CM, i.e., machining. This is the advantage of CM over AM. It allows CM to achieve more efficiency in executing a design than otherwise it would achieve. But this advantage is only design advantage of CM over AM (refer Chap. 5). This advantage is actually a disadvantage if CM is compared wholly with AM. Changing the workpiece orientation decreases the manufacturing speed and causes difficulty to automation. Thus, if inability to change the orientation is a restriction of AM from the design point of view, it is also an advantage from the speed and automation points of view. Due to the speed advantage, AM is called rapid manufacturing. While due to the automation advantage, AM has potential to be direct digital manufacturing (refer Chap. 9).

3 How Does AM Give Freedom to Save Material?

3.1 Principle of Saving Material

AM gives an opportunity to add material where needed. This means only that amount of material is required that will become the constituent of a part. This gives an opportunity to save material. In CM, i.e., machining, a part is machined from a block that is bigger than the part. Therefore, more material is required than that will become the constituent of a part. Thus, AM saves extra material that can otherwise be required in machining.

3.2   Advantage of AM

In CM, i.e., additive processes (sintering, casting. etc.), in principle, materials are not needed more than what is required. Consequently, these conventional additive processes should have the same advantage as that of AM. But these processes do not have the freedom to design as AM has, therefore they cannot utilize the design to save the material as AM can. AM can change the design to change the need where material should be added to save the material—leading to topological optimization, mass optimization, or mass reduction (light-weighting) [8].

3.3   Amount of Material Needed

Though AM fabricates a part by adding the material where needed but it does not imply that AM needs only that amount of material. AM involves more than that amount of material. AM can be classified into two broad categories, i.e., bed process (BP) and deposition process (DP) [9]. Out of these two categories, it is BP that involves more amount of material than DP.

In BP, whether it is powder, slurry, or photopolymer, whole bed needs to be filled with material irrespective of the size, shape, or design of a part to be fabricated. This amount of material is always higher than the amount that will become the part. In photopolymer bed process, a still higher amount of photopolymer is involved if the process is carried out in a container (vat) as the whole container needs to be filled up with liquid irrespective of the size of the part—this situation can be comparable to a particular case of the machining process when a big block instead of a small block is used for machining a small part out.

Thus, for making a part, more extra material is involved than that ends up in the part. If this material can again be utilized in the bed, this will not be wasted. But not all materials that accompany the part in a bed and help develop the part have their properties unchanged. If the material is polymer powder, deterioration rate is high. If the material is metal powder, the deterioration rate is not high. The deterioration rate depends upon the type of material, the size of the part, time spent in a bed system, and scanning time spent by the beam on the bed (refer Chap. 6).

If no material is deteriorated in BP, even then involving so much material in the system does not help completely save the material. Putting the material in the system, recouping the material after the job is done, handling the part and removing the part from the system, cleaning the part [10], cleaning the system—all contribute to varying extent of spilling the material. Thus, involvement of more material does not contribute toward saving the material and is not a factor which gives AM an advantage over CM to save material.

3.4 AM Is Better

Even material deteriorates or spills in BP, there are many cases when AM is better than CM. Considering an example of a titanium part fabricated by machining where only 10% of the material ends up in a part [11] and remaining 90% is either wasted as chips or needed to be recycled [12]. If this part is fabricated in AM, no such high proportion of the material is wasted. Fabricating the same part in PBF may lead some powder not be recycled, especially that powder which is in the vicinity of the part during fabrication, and some powder will be spilled during fabrication. But even then AM saves material. If this example is considered, AM is better than CM concerning saving the material.

3.5 AM Is Not Better

If another example of machining is taken where 50% material ends up in a part and remaining 50% is wasted as chips, AM is not better than CM in this case [13]. It brings a question why this part will be fabricated in AM. This part can still be fabricated in AM if even that small amount of machining is difficult, for example, the material is ceramic or some composite which has less machinability. This brings another question if the part has high instead of less machinability, why the part will be fabricated in AM. The part can still be fabricated if the part is so much complex that even such small amount of machining is difficult to be carried out.

This brings another question if the part has high machinability and the part is not complex, why the part will be fabricated in AM. The part can still be fabricated if the part is just a member of a big group which is going to be fabricated, it is either mass customization or a small batch of production in which decision to fabricate the part in AM is not taken on the basis of merit of AM over CM but due to some logistic reason. In this case, the loss of the material either due to deterioration or spillage is shared by all members of the group, and the disadvantage of AM concerning saving the material concerning fabricating the part is minimized.

What if the fabrication of the part does not accompany a group, the part is one-off, still it can be fabricated if AM furnishes higher mechanical property (refer Chap. 6). Since machining does not improve the mechanical property but AM can and cannot. If AM is giving better properties, the fabrication of the part will not stop only for the reason that there are no other parts ready to be fabricated together and ready to form a group and share material loss.

3.6 Advantage of DP

In DP unlike BP, no bed is formed and thus there is no material consumption by the bed for its own formation. In DP, if material is wasted during fabrication, it is because the deposition of the material was not right, i.e., deposited material was more than that required, or deposited material was not at the right site, or deposited material did not end up being constituent of the part. Material is also wasted in DP when a support structure (SS) is made. But making SS in DP is not a basic requirement of the process in the same way as forming a bed is a basic requirement in BP. Thus in DP, if material is wasted during fabrication of SS, it is not because it has to fulfill some basic requirement as BP has.

Out of two broad categories of AM, it is DP rather than BP has more freedom to save the material when material is wasted for fulfilling the basic requirements of the respective processes. But this basic requirement of BP (which can be a disadvantage if compared to DP and if material is wasted for fulfilling the basic requirement) is also the reason for BP to have an advantage over DP if a complex structure is fabricated.

In AM, it is BP rather than DP which is able to make the most complex structure. DP requires to create more SS than BP for making the same complex structure. It is because in BP, the bed takes majority of responsibility by acting as SS while in DP, there is nothing equivalent to bed which can take such responsibility (refer Chap. 7).

What if all three processes (BP, DP, and machining) and their systems are highly developed. Consequently, in BP, no material is spilled and no material in the bed is deteriorated. In DP, whatever material is deposited ends up in a part and therefore not any material is wasted. In machining, no material is wasted as well since whatever chips are formed and pieces are left after processing, all are well recycled. Then, there will be a question which out of three is better. In this case, DP is the best because whatever material is deposited ended up in a part. BP is not better since it involved more amount of materials for forming the bed than that ended up in a part. Machining is not better as well since it also involved more amount of materials (in the form of extra size of block) for processing to be enabled.

3.7 How Material Is Wasted

Material is wasted in AM as follows:

1. Wastage of feedstock during fabrication,
2. Wastage of feedstock due to the lack of its processability,
3. Spillage of feedstock,
4. Wastage of processed feedstock due to some error during fabrication,
5. Wastage due to the formation of SS, sizing of oversized parts, finishing of rough parts, etc.

3.8   How Material Is Saved

Material is saved in AM as follows:

1. Recycling of feedstock,
2. Reuse of feedstock for fabrication,
3. Using one material in one machine to minimize spillage and contamination,
4. Recycling of processed feedstock,
5. Making many parts simultaneously,
6. Development and use of small AM system for investigation,
7. Monitoring the fabrication and stopping the processing in case of any failure so that material to be processed after the failure can be saved,
8. Changing the orientation to do away with the need for making SS,
9. Fabrication of complex parts in BP rather than in DP to minimize wastage due to SS,
10. Making SS in not different materials in DP in order to be able to recycle or reuse processed material. If two materials (one for SS and other for part) are used, it is possible both materials will mix during fabrication. Their mixing will decrease their recyclability and reuse (refer Chap. 7).

4   How Does AM Give Freedom to Have Many Properties?

From the same material, AM gives opportunity to develop a number of products having different properties. Since properties depend upon how processing parameters are set—by changing parameters different properties are achieved. These properties will be suitable for different applications and thus the same material can be used for developing different products. For example, if a high power beam is applied in PBF, melting will occur which will give rise to high strength. The resulting product can be used in a critical application such as implants. If a lower power beam instead is applied so that partial melting instead of full melting occurs, it will give rise to low strength and porosity. The resulting product cannot be used in critical application where uncertainty in product properties can lead to disastrous effects. The product though can be used in non-critical application such as fixtures to hold workpiece in machining.

4.1   By Changing Parameter

If AM is compared with CM, i.e., machining, it is clear that this type of opportunity does not exist in machining where due to different machining parameters different products can be fabricated for different applications. But if AM is compared with additive process based CM such as sintering, casting, then CM is also capable to make different types of products by changing different experimental parameters such as temperature, pressure, etc.

But AM gives opportunities to change the properties at different locations of a product by changing the parameters. For example, if higher beam energy is applied at the last layer of an unfinished product to cause melting, the outer surface of the product which is made by the last layer will have higher strength than the inner layers which are made by partial melting applied by lower beam energy.

Thus a product can be designed to have one property at its outer surface and other property at its center. If the processing parameters are optimized and are gradually changed from the center to the surface, a gradual change in properties in a product can be achieved. This will allow a product to have many properties at its different locations to serve different functions without changing the material. Changing the parameters will create a degree of porosity which can be used for different applications [14].

CM does not allow to achieve such variation in properties without changing the material. In sintering, some change in properties can be achieved by subjecting the mold to different parameters at its different sections but if design is complex to precisely control the parameters and resulting properties, changing the parameters will not be as convenient as in AM. In casting, the properties of a product at its different areas can be varied by changing the flow rate or interrupting the flow and changing the temperature of the mold, etc. But the variety in properties that will be achieved in casting will not be the same as in AM.

4.2   By Changing Material

If AM gives opportunities to vary properties in a product without changing the material, what will happen if AM acquires facility to change material or feedstock.

DP instead of BP allows to change material conveniently [15]. Multi-material products can be formed in DP if there is a provision to incorporate feedstock of additional materials. For example, in wire DP if there is a provision to incorporate many wires each made of different materials, or similarly if there is a provision for many filaments or powders or jetted drops [1] in filament DP or powder DP or material jetting, respectively.

Consider laser powder deposition system which is equipped with two powder feeders each is connected with different powder reservoirs. By assigning different feed rates to feeders, different amounts of each powder can be fed which will enable to make a number of two-material components containing contributions from different amounts of each powder. Each component can have different properties and thus AM will allow to make a number of products having different properties using two materials which is not possible in CM [16] (refer Chap. 5).

But it requires a number of investigations:

- Materials should not have high difference in vapor pressure otherwise one will melt while other will evaporate bringing an undesirable change in final composition of the component, e.g., Al vaporizes during the processing Ti-6Al-4 V [17],
- Powder size of two materials should not be much different otherwise one powder will not be melted while other will evaporate,
- Materials should not have different coefficients of thermal expansion; otherwise crack will be generated due to stress [18],
- Materials should not react and change their chemical states,
- Material must not make deleterious intermetallic compounds. This can be avoided in DP by using another material between two materials. Thus, layer made by another material will become an intermediate layer between two layers made from two different materials. Intermediate layer will help overcome difficulties arising due to different thermal expansion and solubility (of materials) between two layers [19] (Fig. 2.4).

4.3 Due to High Cooling Rate

High rate of solidification furnishes unique opportunities such as those materials which are mutually incompatible in equilibrium condition and are immiscible in CM can make unique compound defying the equilibrium state. Due to the non-equilibrium condition created by high solidification rate, these immiscible materials will solidify before they are segregated. This gives opportunity not only to mix materials and make a component having new property but to use AM as a tool to discover new material [20].

Checking material phase diagram, it will be clear how many compounds are possible to be formed. Since the phase diagram represents equilibrium conditions at various temperatures, there are few compounds possible to exist at room temperature. Though there are many compositions possible to exist at high temperature. However, if these compositions are allowed to cool slowly, by the time they reach room temperature, their compositions change and acquire the composition corresponding to the room temperature.

The only way to retain the compositions at high temperature is to not let them get transformed by not allowing them to cool slowly because when they cool slowly they get sufficient time to get transformed. It is possible by increasing the cooling rate so that enough solid state diffusion should not occur. There are many AM processes which work with high rates of cooling and thus give the desired condition.

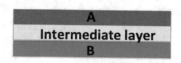

Fig. 2.4 Use of intermediate layer to increase bond between two layers made by different materials (**a** and **b**)

Though, high cooling rate gives cracks and defects which need to be prevented or suppressed using preheating or optimizing parameters or alloying with nanoparticles [21].

The desired condition allows the composition which otherwise exists only in a liquid or semi-liquid state also to exist at room temperature in a solid state—this is how a process allows to freeze the composition. As these compositions are forcefully brought to the solid state, which is their non-equilibrium state, they will leave the solid state if given the opportunity (such as heating). In the absence of such opportunity, their transformation is too slow to change the state. Thus, AM gives a number of compositions to be translated into products. Though, the products will have residual stresses due to the high rate.

This has helped make better high entropy alloy by grain refinement which was not possible by CM [22]. Though, high rate of solidification does not allow to form all compounds [23] that can be possible by changing ratio of materials. This is because high rate has itself limitation. This limitation is not due to unavailability of high power beam or high scan rate but due to rate of heat transfer and diffusion of material enabling solidification.

Besides metals, plastic parts are also found to have better mechanical properties than CM parts [24].

4.4   By Changing Deposition in Different Directions

What if powder ratio is varied not only in the build direction (z-axis) but in x- and y-directions as well, many two-material products each having many properties distributed at various locations can be formed. When instead of changing only powder ratio, number of nozzles depositing different powders can be increased, higher varieties of multi-material products can be fabricated. Thus, a number of materials can be developed and a number of products can be fabricated if full potential of AM is realized. Similar to powder, other feedstocks such as wire or photopolymer have potential to bring variations in different directions.

5   How Does AM Give Freedom to Move?

5.1   Advantage of AM System

AM is a self-contained system. To make a useful product, it does not need many sub-processes as machining needs, or it does not be like an injection molding machine which has limited scope in limited time, or it is not similar to forging which relies on heavy tooling making the forging nonportable. Thus, AM gives opportunity to have a system which is compact, which can be placed anywhere to be exploited, and which is more useful and versatile than any single contemporary manufacturing system. It can be placed in a strategic location where a number of other machines are required to do the same job which it will do single-handedly, e.g., for providing automotive spare parts by being present somewhere in an automotive supply chain. An AM system can be carried to a space station where it can be used to make emergency parts; for this application there is no alternative to AM.

If AM is used, manufacturing is no longer having restriction to be carried out in a designated space and the system is no longer inoperable by a non-specialist not well-versed in programming. Henceforth, a number of parts can be fabricated using one machine in a home environment. Earlier it was not possible, there could be a lathe machine in a home workshop to make limited type of metallic parts, and there were no systems available to make plastic parts. AM has allowed, in one extreme, creation of a number of hobbyist parts more than that required, and in other extreme, fabrication of an emergency part for a broken equipment without which further use of the equipment can be in jeopardy.

5.2   Industrial Implication of AM System

The ability of an AM machine both to be located somewhere and to function well can have industrial implication. Imagining a state when a part is needed at a place but the part cannot reach there because the place is in a remote area, or the spare part is not available to move on, or the spare part is available but identifying the part out of thousand parts is not a day's task, or transporting the part will take considerable time—if there will be an AM system which can be located anywhere not far away from the place of requirement, it can supply such parts and will be a solution to many problems.

These problems are transportation cost, amassing inventories (a number of spare parts for future need), transportation delay, keeping tracks of thousand parts in a storehouse, deterioration of parts with the passage of time or due to lack of upkeep [25], turning of the stored part into waste because before the part is used in an old equipment a new upgraded equipment is invented which does not need this part anymore, etc.

5.3   Limitation of AM System

But, whether an AM system can make all parts required; it can make a number of parts but not all parts. AM is using a number of materials but the parts that are formed by CM and are stored as spare parts are made in many other materials as well for which AM has no experience. There are many spare parts for which AM is not certified and the properties are not guaranteed. The standardization in AM is an ongoing development and is not as mature as in CM. The automation in AM is not as developed as in CM (refer Chap. 9).

Thus an AM system placed at a strategic location of supply chain network has limited solutions to offer. It cannot make all spare parts needed. In order to increase the capacity at the strategic location, one AM system may not be sufficient, some other AM systems are required which will work with different materials such as plastics and ceramics. More than one metallic AM system may be required as well for working with various metallic alloys—it will help speed up the fabrication speed, avoid contamination with other metals, some design works better in one AM system than in the other system. For example, a cylindrical part can be economically fabricated in DP than in BP.

One AM system will not work if the aim is to provide all required spare parts. Many systems may be required instead. It means one room containing one system will not help, a bigger place containing many systems will instead suffice. The place will then resemble like a factory. But, it is not expected from AM to provide a factory in a supply chain. If AM solves the problem by creating such factory, AM is not providing solution but is replacing one problem with another problem. When the problem becomes bigger, AM does not provide solution. It is not expected. There is a dream that there will be one or two systems which will be sufficient for all problems in a supply chain. But, if such a new factory is required, the dream is shattered. Unless AM is developed, it has limited application as a spare part supplier [26].

6   How Does AM Give Opportunity to Save Energy?

AM, in general, is an energy intensive process which is a disadvantage and thus is not able to save energy by process itself. But AM is useful to save energy indirectly by its applications.

6.1   By Repair

AM can repair, refurbish, or remanufacture a damaged part and thus AM helps the part not be lost but come back in use [27]. For example, in the absence of repair of a damaged turbine blade, it cannot be reused and the energy spent on its fabrication will be lost. Thus repair saves that energy. If the damaged part is not recycled, the material of the part will again be lost. With this loss, the energy spent in producing that material is again lost. AM thus, by virtue of repairing, saves energy associated with the fabrication of a damaged part and with the production of its material [28] (refer Chap. 6).

If a part is successfully repaired by CM, AM is not having any advantage over CM; AM is then just another tool like CM to save energy by repair. But if the part cannot be repaired by CM but by AM, AM is not just another tool like CM but the only tool to save energy by repair.

6.2   By Energy-Efficient Part

AM is able to make an efficient part that was not possible earlier. When these parts are used they last long and work efficiently and thus saves energy. For example, a turbine blade having internal cooling channel lasts longer because cooling decreases its probability of getting damaged by taking the heat away (refer Chap. 6). Thus longevity of a part is improved leading to a decreased need of reproduction of the part, and thus saving the energy required for reproduction.

6.3   By Light-Weighting

AM is able to make light weight parts such as automotive or aerospace parts which when used saves fuel energy because being lighter in weight the vehicle needs less fuel [29]. But, if the light part is not used as automotive or aerospace parts, AM does not help save energy [13] (refer Chap. 6).

6.4   By Assembly

If many parts are fabricated by CM and they are assembled by energy intensive process, the energy spent in assembly can be saved if a single part consisting of all parts is fabricated. AM gives an opportunity to save that energy (for assembly) by fabricating a single part (instead of separated many parts) [30] that cannot be fabricated by CM.

But AM will only be able to save that energy if for fabricating that single part requires less energy than the combined energy (energy spent for fabricating in CM and energy spent for assembly). Thus if a product consists of many parts, it saves energy. Example, General Electric replaced 20 cast parts by a single AM part that was less expensive and more durable [31].

Figure 2.5 shows the concept of assembly where seven parts are assembled in one part by AM. In the absence of AM, these parts can be made by some other processes such as casting or machining (Fig. 2.2a), or can be assembled by welding (Fig. 2.6b). Thus AM is a one-step process represented by Fig. 2.1 while CM is a two-step process represented by Fig. 2.6a, b, respectively. It can be argued that complex part as shown in Fig. 2.5a can also be fabricated by casting or machining,

Fig. 2.5 Assembled
product by AM

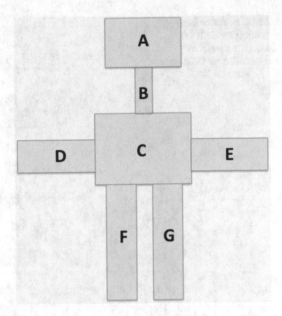

but if the part becomes complex, CM will not be able to provide the ease of fabrication that AM provides.

Similarly, Fig. 2.7 shows an interlocked part that can be made by AM. If this part needs to be made by CM, one of the rings will be made by machining, casting, or forging [32] (Fig. 2.8a) while other ring will be connected to it by cutting and welding (Fig. 2.8b).

6.5   By Creating Energy

AM helps develop better energy materials and makes parts for devices used in energy production such as fabrication of battery components, and thus contributes to creating energy [33]. Though energy creating is not equal to energy saving but if seen from bigger perspective, a process which wastes energy but is able to create energy as well, its demerit of wasting energy is diluted.

Fig. 2.6 Assembled product by CM: (**a**) separate parts, (**b**) assembled by welding

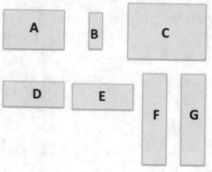

(a) Separated parts made by machining or casting

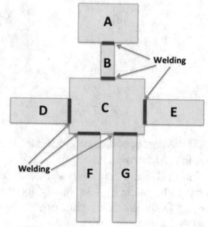

(b) Parts assembled by welding

Fig. 2.7 Interlocked part consisted of two rings made by AM

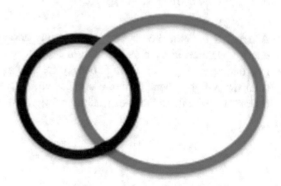

Fig. 2.8 Interlocked part assembled by CM: (**a**) separate metallic rings made by casting and machining, (**b**) separate rings are interlocked by welding

(a) separate metallic rings made by casting and machining

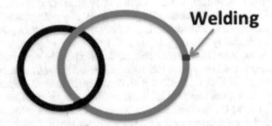

Welding

(b) Separate rings are interlocked by welding

References

1. Yuan, C., Wang, F., Rosen, D. W., & Ge, Q. (2021). Voxel design of additively manufactured digital material with customized thermomechanical properties. *Materials and Design, 197*, 109205.
2. Korpela, M., Riikonen, N., Piili, H., et al. (2020). Additive manufacturing—Past, present, and the future. In M. Collan & K. E. Michelsen (Eds.), *Technical, economic and societal effects of manufacturing 4.0*. Cham: Palgrave Macmillan.
3. Hällgren, S., Pejryd, L., & Ekengren, J. (2016). Additive manufacturing and high speed machining -cost comparison of short Lead time manufacturing methods. *Procedia CIRP, 50*, 384–389.
4. Khoda, B., Ahsan, A. M. M. N., Shovon, A. N., et al. (2021). 3D metal lattice structure manufacturing with continuous rods. *Scientific Reports, 11*, 434.
5. Jin, Q. Y., Yu, J. H., Ha, K. S., et al. (2021). Multi-dimensional lattices design for ultrahigh specific strength metallic structure in additive manufacturing. *Materials and Design, 201*, 109479.
6. Gu, H., Li, S., Pavier, M., et al. (2019). Fracture of three-dimensional lattices manufactured by selective laser melting. *International Journal of Solids and Structures, 180-181*, 147–159.
7. Carneiro, V. H., Rawson, S. D., Puga, H., et al. (2020). Additive manufacturing assisted investment casting: A low-cost method to fabricate periodic metallic cellular lattices. *Additive Manufacturing, 33*, 101085.
8. Zhu, L., Li, N., & Childs, P. R. N. (2018). Light-weighting in aerospace component and system design. *Propulsion and Power Research, 7*(2), 103–119.
9. Kumar, S. (2020). *Additive manufacturing processes*. Cham: Springer.
10. Verhaagen, B., Zanderink, T., & Rivas, D. F. (2016). Ultrasonic cleaning of 3D printed objects and cleaning challenge devices. *Applied Acoustics, 103B*, 172–181.

11. Dominguez, L. A., Xu, F., Shokrani, A., et al. (2020). Guidelines when considering pre & post processing of large metal additive manufactured parts. *Procedia Manufacturing, 51*, 684–691.
12. Razumov, N. G., Masaylo, D. V., & Silin, A. O. (2021). Investigation of additive manufacturing from the heat-resistant steel powder produced by recycling of the machining chips. *Journal of Manufacturing Processes, 64*, 1070–1076.
13. Ingarao, G., Priarone, P. C., Deng, Y., & Paraskevas, D. (2018). Environmental modelling of aluminium based components manufacturing routes: Additive manufacturing versus machining versus forming. *Journal of Cleaner Production, 176*, 261–275.
14. Attar, H., Löber, L., Funk, A., et al. (2015). Mechanical behavior of porous commercially pure Ti and Ti–TiB composite materials manufactured by selective laser melting. *Materials Science and Engineering A, 625*, 350–356.
15. Laitinen, V., Merabtene, M., Stevens, E., et al. (2020). Additive manufacturing from the point of view of materials research. In M. Collan & K. E. Michelsen (Eds.), *Technical, economic and societal effects of manufacturing 4.0*. Cham: Palgrave Macmillan.
16. Kelly, J. P., Elmer, J. W., Ryerson, F. J., et al. (2021). Directed energy deposition additive manufacturing of functionally graded Al–W composites. *Additive Manufacturing, 39*, 101845.
17. Mukherjee, T., Zuback, J., De, A., & DebRoy, T. (2016). Printability of alloys for additive manufacturing. *Scientific Reports, 6*, 19717.
18. Wei, K., Xiao, X., Chen, J., et al. (2021). Additively manufactured bi-material metamaterial to program a wide range of thermal expansion. *Materials and Design, 198*, 109343.
19. Zhang, X., Sun, C., Pan, T., et al. (2020). Additive manufacturing of copper – H13 tool steel bi-metallic structures via Ni-based multi-interlayer. *Additive Manufacturing, 36*, 101474.
20. Hofmann, D., Roberts, S., Otis, R., et al. (2014). Developing gradient metal alloys through radial deposition additive manufacturing. *Scientific Reports, 4*, 5357.
21. Xue, J., Feng, Z., Tang, J., et al. (2021). Selective laser melting additive manufacturing of tungsten with niobium alloying: Microstructure and suppression mechanism of microcracks. *Journal of Alloys and Compounds, 874*, 159879.
22. Zhao, W., Han, J. K., Kuzminova, Y. O., et al. (2021). Significance of grain refinement on micro-mechanical properties and structures of additively-manufactured CoCrFeNi high-entropy alloy. *Materials Science and Engineering A, 807*, 140898.
23. Martin, J., Yahata, B., Hundley, J., et al. (2017). 3D printing of high-strength aluminium alloys. *Nature, 549*, 365–369.
24. Aretxabaleta, M., Xepapadeas, A. B., Poets, C. F., et al. (2021). Comparison of additive and subtractive CAD/CAM materials for their potential use as Tübingen palatal plate: An in-vitro study on flexural strength. *Additive Manufacturing, 37*, 101693.
25. Ford, S., & Despeisse, M. (2016). Additive manufacturing and sustainability: An exploratory study of the advantages and challenges. *Journal of Cleaner Production, 137*, 1573–1587.
26. Chekurov, S., Metsä-Kortelainen, S., Salmi, M., et al. (2018). The perceived value of additively manufactured digital spare parts in industry: An empirical investigation. *International Journal of Production Economics, 205*, 87–97.
27. Walachowicz, F., Bernsdorf, I., Papenfuss, U., et al. (2017). Comparative energy, resource and recycling lifecycle analysis of the industrial repair process of gas turbine burners using conventional machining and additive manufacturing. *Journal of Industrial Ecology, 21*, S203–S215.
28. Wilson, J. M., Piya, C., Shin, Y. C., et al. (2014). Remanufacturing of turbine blades by laser direct deposition with its energy and environmental impact analysis. *Journal of Cleaner Production, 80*, 170–178.
29. Verhoef, L. A., Budde, B. W., Chockalingam, C., et al. (2018). The effect of additive manufacturing on global energy demand: An assessment using a bottom-up approach. *Energy Policy, 112*, 349–360.
30. Boschetto, A., Bottini, L., Eugeni, M., et al. (2019). Selective laser melting of a 1U CubeSat structure. Design for additive manufacturing and assembly. *Acta Astronautica, 159*, 377–384.
31. Niaki, M. K., Torabi, S. A., & Nonino, F. (2019). Why manufacturers adopt additive manufacturing technologies: The role of sustainability. *Journal of Cleaner Production, 222*, 381–392.

32. Sun, M., Xu, B., Xie, B., et al. (2021). Leading manufacture of the large-scale weldless stainless steel forging ring: Innovative approach by the multilayer hot-compression bonding technology. *Journal of Materials Science and Technology, 71*, 84–86.
33. Zhakeyev, A., Wang, P., Zhang, L., et al. (2017). Additive manufacturing: Unlocking the evolution of energy materials. *Advancement of Science, 4*, 1700187.

Chapter 3
Disadvantage

1 What Are the Disadvantages of AM?

Following are four disadvantages of AM (Fig. 3.1):

1. Slow process
2. Energy intensive process
3. Lack of repeatability
4. Lack of processable materials

2 How Is AM a Slow Process?

2.1 Why Machining Is Fast

For converting material into a part, AM takes more time than CM. For example, machining a part of size 10 cm^3 from a block will take less time than building a part of the same size in AM from small blocks or entities such as powder, wire, or liquid drop. In machining, a part is made faster because machine tool cuts the block at contour to carve out a part. During this cutting, majority of the material of the block remains untouched by the cutting edge of the tool. This results in majority of the

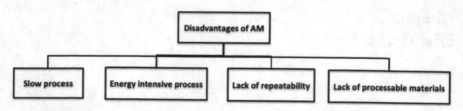

Fig. 3.1 Disadvantages of AM

material of the part remaining uncut by cutting. Thus after machining, a lot of volume of the part does not go through the experience of interacting with a machining tool but still remains the constituent of a part which even though is made by machining.

2.2   Why AM Is Slow

In AM, a part is made by adding small materials and the part is the result of having majority or all of the materials remain affected by joining. Even a small feature of an AM part is not found to be fabricated by not going through joining a number of small blocks or entities. Since AM unlike CM does not have freedom to have a part made by leaving some volume of the part remains unprocessed, it is not able to save time and labor invested in processing that volume which can otherwise be saved in a machining process. This requirement of extra time makes AM slower than machining.

2.3   When AM Is Fast

AM can be faster if scanning (for processing materials) done in AM is so fast that even if scanning needs to cover a far longer path (than a cutting tool path in machining), the scanning takes less time than the cutting. AM can also be faster even if scanning done in AM is not faster if cutting tool is not able to move fast because cutting tool path is too complex, and though cutting tool has to cover a smaller path it is not able to manage to cover fast. Thus the disadvantage of AM can be overcome by making a complex part that will take more time than a simple part to be machined, while the same part will not take such more time in AM.

Thus the slowness of AM will not be visible in parts of all types of geometries.

3 Why Is AM an Energy Intensive Process?

AM makes parts by joining, the act of joining requires energy. Since there is a lot of small pieces to be joined, it requires more energy. If the problem was only simple joining, it might not require so much energy—all pieces might be brought together and the heat might be applied. This happens in conventional sintering. But AM does not do joining in that way. AM joins pieces in a selective way, while it joins some pieces other pieces wait to be joined. Considering an example for making a part from metal powders in PBF, while heat is applied to join some pieces, some heat gets squandered that are not utilized to join other pieces which are waiting.

There is not yet any PBF available where such heat is not squandered. How much heat is squandered depends upon the efficiency of the process—this squandering is due to the lack of absorption of heat, over-absorption, absorption by other pieces, etc. This is the side effect of joining materials in a bit by bit fashion that is AM. If they were joined together as in conventional sintering, energy could not be lost in that way. Thus AM requires more amount of energy to do the same amount of joining, if compared to CM—it makes AM energy consuming or energy intensive process. For example, for making a part from steel powder, AM consumes 97 MJ/kg while conventional sintering consumes only 2.4 MJ/kg [1].

3.1 There Is No Shortcut

The best way to stop wastage of heat is not to use heat. Considering the same case of joining of metal powder when metal powders are not joined by melting but are glued using some ink or adhesive binder [2]-in this case, heat or energy is not required to change their metallurgical states in order to join them and thus there is no opportunity for heat to be applied and then get wasted. There are AM processes which join without applying high energy and therefore there is less chance for wastage of energy, but these processes do not make strong parts.

If parts made using adhesive binder need to be made stronger, they are again treated in a furnace to go through sintering. This is a post-processing (of AM parts) which requires energy. Thus the fabrication in some AM requires energy twice: one during fabrication and the other during sintering. If the part is made by conventional sintering, the fabrication requires energy just once during sintering. Though the energy required in post-processing sintering and conventional sintering is not the same. But the energy comparison gives a glimpse that taking help of post-processing does not imply less investment in energy for the fabrication of a high-strength part through AM route. Thus it is not possible to bypass the requirement of energy for making a high-strength part by avoiding more energy intensive AM process (and instead choosing less energy intensive AM process) and compensating the loss (in properties) by using post-processing.

3.2 It Is Not Always Demerit

If AM is energy intensive, it does not imply that it is not more energy efficient than CM. Being more energy intensive than CM only implies that for processing a given area, AM will require more energy than CM. But what if making a part having the same size and design does not lead the same area to be processed in AM and CM. For making the part, different areas need to be processed in AM and CM since processing in AM is different than in CM.

What if the area needed to be processed in AM is significantly less than in CM for making the same part—so that total energy spent in AM is less than in CM. Considering a case of machining of a titanium part from a titanium workpiece where only 2% of the workpiece is left (as a part) and 98% of the workpiece needs to be removed chip by chip. If the same part is made in AM, a small area (roughly equivalent to an area represented by only 2% of the workpiece used in CM) needs to be processed and a small energy will be spent. This titanium part is a complex part and therefore making a complex part by AM will save the energy in comparison to CM.

If a part is made in AM and addition of materials does not lead to enough complexity in the final part, a simple part forms which can be easily made in machining. In this case, the energy spent in machining is small while the energy spent in AM is high. If again a part is made in AM and addition of materials leads to enough complexity in the final part, a complex part forms which cannot be easily machined. In this case, energy spent in machining is high while the energy spent in AM is almost the same as in previous case. Though AM is an energy consuming process, the consumption does not increase as the complexity of the part increases—but the consumption increases in machining process. This consumption in AM is not much when the consumption in machining is also much. This demerit of AM is not prominent when the merit of the other process (machining) fails to deliver when complexity starts to arrive.

It can be argued that a complex part when will be formed in AM will require support structure (SS), and the energy spent in fabricating the part will be equal to the energy spent in fabricating the part plus the energy spent in fabricating SS. Consequently, the energy spent becomes more due to SS. But the fabrication of SS is optional and depends upon the type of material, design, and strategy to manufacture. In an ideal case, it is assumed that the process is developed well with respect to a material or a design or a strategy that no SS is required. There can be a worst case scenario when AM is not having energy advantage over machining due to the additional energy spent in SS. But this scenario does not negate the fact that with an increase in complexity while machining moves towards consuming more energy AM does not move towards consuming more energy.

3.3 Energy Sharing

Making parts using machining will require less energy than that required in AM. If machining is a fine machining, energy required will more than that of rough machining since more time is spent in material removal.

Energy spent for fabricating a part is high in AM but if hundred parts can be accommodated in an AM system, total energy spent will be shared by each part. Consequently, fabrication of hundred parts requires few energy per part. Whether this part will require more energy than a fine machined part will depend upon how much complex the part is and how many AM parts are fabricated.

4 What Is the Lack of Repeatability in AM?

4.1 Meaning of Repeatability

Repeatability means two parts will not be different if they are made in any of the following ways:

- Using the same AM process [3],
- Using the same feedstock, parameters, and machine,
- Using the same process but different machines [4, 5],
- Using the same machine (from the same company) but different models,
- By different operators in the same machine,
- At different locations inside a machine [6, 7],
- At different times,
- In two different batches, etc.

If repeatability is limited only to dimension, the parts after fabrication can be reworked [8], but if mechanical properties are different, reworking will not work.

In AM, a part is fabricated by joining material blocks. Since the size of the part is relatively big in comparison to the material block size, there are many bonds needed to be formed to make a part. If two identical parts need to be fabricated, the bonds in two parts must be similar. If the bonds are different, their difference must not change the part properties in a significant way so that two parts will no longer be considered identical.

4.2   Repeatability in CM

Formation of a part from blocks or drops is not new in CM, it happens in conventional sintering or casting, respectively, where property of a part depends on many bonds that from between blocks or drops. These bonds drive the properties of parts thus formed.

In CM, all materials or feedstocks that constitute a part are subjected to the same external condition. For example, if a part is made in conventional sintering, the mold is kept at the same temperature and pressure which implies that fabrication of a part relies on having all powder to have the same external condition. Having one external condition for all powder may not ensure the same intended effects on all powder occupying different myriad space of the mold. But conventional sintering unlike AM does not bring intended effects on all powder by applying several space-specific external conditions.

The application of an external condition gives rise to various types of bonds among particles confined within the mold. For making the next part, the same external condition is applied. It may be possible that the same condition does not give rise to the same type of bonding in two successive parts and their properties may vary causing no repeatability. This can be due to unintended variations in temperature, pressure, or change in mold condition for the next part. But these variations are due to one external condition applied to whole bunch of powder together, and therefore if one external condition is controlled or if measures are taken which ensure that these variations do not originate, there can again be repeatability and two parts made in CM will be identical in properties.

But what if there is no single external condition to be managed but there are many external conditions to be managed, there is more possibility that repeatability will not be ensured.

4.3   Lack of Repeatability in AM

In AM, for example in PBF, all powder belonging to a part are not treated once but they are laid on as several layers. Each layer is divided into several lines to be treated separately. When one line is treated to convert it into some section of a part, the line is subjected to an external condition which is set in the form of optimized parameters. All lines in a layer are treated with the same optimized parameters and are thus subjected to the same external condition. But, this condition is applied separately at different lines at different times, there is no guarantee that the effect of the same condition will not be different at different lines in the same layer [9].

If it is assumed that the total effect of same parameters (external condition) on all similar lines in a layer is equal to the total effect of all similar lines in another similar layer, it does not imply that the total effects of all similar layers of the first section of a homogeneous part will be equal to the total effects of all similar layers of the next section of the same part. Consequently, even after so many assumptions, there is no guarantee that there will be repeatability from one part to the next part. In order to ensure repeatability, each and every conditions need to be duplicated when next part is formed. Since a part is made up of several layers of different sizes and locations and a layer is made up of several lines of different sizes and orientations, there are far more external conditions needed to be controlled and managed than if the part is made up of just one bunch (as in CM). Thus there is a lack of repeatability in AM while there is more repeatability in CM.

4.4 Comparison

In CM, i.e., machining, a part is made by cutting. The cutting separates a piece from a workpiece, the piece is called a part. This act of separation is confined on the surface and does not work inside the part. Depending upon the type of cutting, the surface can have low or high roughness or oxide layer, etc. If the cutting is severe, the damage from the surface can reach to the bulk in the form of overheating or bending. If there is overheating, there will be heat-induced material transformation. Even in the worst case scenario, the bulk material of a part is the last to get damaged in machining. If there is a problem with repeatability in machining, it is not because machining changes the bulk material property in one part while it does not change the same in the next part.

While in CM, the act of separation works on surface with minimal prospect to change the bulk, there is no such guarantee in AM. On the contrary, in AM the act of joining though starts from the surface but it is not certain that joining will be confined within the surfaces of blocks (powder, wire). In many AM processes, the joining transforms the whole block (surface + bulk) from solid to liquid and then to solid again. With this transformation, it is not certain that feedstock (powder, wire) property will be continued after the joining.

In CM, there is no such transformation and there is almost guarantee that the feedstock (workpiece) property will be continued after machining. There was a feedstock property before machining and there is a feedstock property after machining and this feedstock property is a part property. In AM, there was a feedstock property before joining and there is no feedstock left anymore to have any property and there is a part property which is not the feedstock property.

While in some other AM processes where there is no such solid–liquid–solid transformation, there was a feedstock property before joining and there is a feedstock property after joining and there is a part property which is due to both feedstock and joining. Since joining is a net sum of the effect of number of small joining corresponding to a number of small feedstock blocks, joining may have many values depending upon the type of small joining.

In order to have repeatability in AM, small joining must have the same value in all parts giving rise to same property in all parts while there is no corresponding small cuttings in CM and repeatability in CM is free from such requirements. Therefore, there is more repeatability in CM while there is less repeatability in AM.

5 Why Is There Lack of Processable Materials in AM?

AM does not have many materials for industrial applications. AM has approximately 29 metallic powders [10–12], 15 metallic wire, five polymer powders, five polymer filaments, and five photopolymer resins. These numbers are miniscule in comparison to several thousand materials [13, 14] available in CM.

There are hundred of commercial metallic materials having different trade names available in AM [15]. A total of 129 polymer materials including 66 for powder bed process are listed [16] which gives an impression that there is not so low number of processable materials in AM. But these commercial materials are either same basic AM materials or developed from basic AM materials for a specific application or machine. Thus these customized materials do not increase the actual number of basic AM materials.

There are following six reasons why there is few processable materials in AM in comparison to CM:

1. AM is a new process, it does not have a history of more than 40 years while CM has a history of more than 170 years (counting from 1850 when industrial revolution started).
2. One of the applications of AM is prototyping for which having many materials is not a necessity. Therefore, prototyping is not the application that demands new materials and is not a motivating factor for creating new materials.
3. AM has design advantage over CM—this has given rise to new products which were earlier not possible, this has again potential to create many more products. Most of the AM research goes in this direction to exploit the design advantage of AM. For this purpose, creating new materials is not required as much as the testing of new design with existing materials. Creating a new product using a new design does not get setback due to the lack of materials, and thus there is no motivation for creating new materials from the design point of view. It is not surprising that same polyamide is getting used to create chairs as well as human implants.

It does not mean existing number of materials are sufficient, or new types of products will not be needed or formed if a new material comes. It only means existing materials are not insufficient that still new products cannot be made.

1. In AM, a material has to fulfill higher requirement than in CM to be a successful material. In CM additive processes such as sintering or injection molding, materials just need to coalesce while they get help from a mold to give rise to a shape. In AM, materials need to coalesce in such a way that it will lead to a shape [17]. There is no help from any mold, the onus is on the material to play double role and do the duty of a mold. While there is only one requirement in CM to be a successful material, there are two requirements in AM to be equally successful. Thus, whatever is the requirement in CM to be a successful material is not sufficient to be successful in AM.
2. In AM, standard is high while in CM, such high standard is not required for the acceptance of a material. For example, in machining, aluminum alloys have

machinability. With a slight change in percentage of alloying element such as Mg or Si, machinability does not change drastically. Thus, there is a range of aluminum alloys which are considered to have acceptable machinability. A slight variation in machinability due to change of alloying element does not make the material unfit to be accepted for machining. In AM, with a slight variation in Mg or Si, if the aluminum alloy shows an increase in porosity, this variation in property may not make the material acceptable. Thus due to high standard in AM, there is a lack of materials.

3. There is a large variation among AM processes and the material developed for one AM process is not transferrable to another AM process of the similar type. Therefore, each process requires process-specific attention for developing a material. For example, a powder that works in PBF has no guarantee that it will work in powder deposition process, or a powder that works in LPBF has no guarantee that it will work in EPBF. Photopolymer resin developed for photopolymer bed process has no guarantee to work in photopolymer jetting process.

While there is no such large variation among CM processes, if a material works in milling process, the material can also work in turning or drilling. If a material such as aluminum has high ductility and will work in a forming process, the material can also work in other process, i.e., forging. Therefore, variation among CM processes has not as much impact on CM material development as the variation among AM processes on AM material development.

References

1. Azevedo, J. M. C., Serrenho, A. C., & Allwood, J. M. (2018). Energy and material efficiency of steel powder metallurgy. *Powder Technology, 328*, 329–336.
2. Bai, Y., & Williams, C. B. (2018). Binder jetting additive manufacturing with a particle-free metal ink as a binder precursor. *Materials and Design, 147*, 146–156.
3. Moylan, S., Slotwinski, J., & Cooke, A. (2012). Proposal for a standardized test artifact for additive manufacturing machines and processes. In *SFF Proceedings* (pp. 6–8).
4. Vorkapic, N., Pjevic, M., Popovic, M., et al. (2020). An additive manufacturing benchmark artifact and deviation measurement method. *Journal of Mechanical Science and Technology, 34*, 3015–3026.
5. McGregor, D. J., Rylowicz, S., Brenzel, A., et al. (2021). Analyzing part accuracy and sources of variability for additively manufactured lattice parts made on multiple printers. *Additive Manufacturing, 40*, 101924.
6. Seifi, M., Christiansen, D., Beuth, JL., et al. (2016). Process mapping, fracture and fatigue behavior of Ti-6Al-4V produced by EBM additive manufacturing. In *Proceedings of 13th World Conference on Titan* (pp. 1373–1377).
7. Galarraga, H., Lados, D. A., Dehoff, R. R., et al. (2016). Effects of the microstructure and porosity on properties of Ti-6Al-4V ELI alloy fabricated by electron beam melting (EBM). *Additive Manufacturing, 10*, 47–57.
8. Shah, P., Racasan, R., & Bills, P. (2016). Comparison of different additive manufacturing methods using computed tomography. *Case Studies Nondes Test Eva, 6*, 69–78.
9. Dowling, L., Kennedy, J., O'Shaughnessy, S., & Trimble, D. (2020). A review of critical repeatability and reproducibility issues in powder bed fusion. *Materials and Design, 186*, 108346.

10. Qian, M. (2015). Metal powder for additive manufacturing. *JOM, 67*, 536–537.
11. Zhang, L. C., & Attar, H. (2016). Selective laser melting of titanium alloys and titanium matrix composites for biomedical applications: A review. *Advanced Engineering Materials, 18*, 463–475.
12. Anderson, I. E., White, E. M. H., & Dehoff, R. (2018). Feedstock powder processing research needs for additive manufacturing development. *Current Opinion in Solid State & Materials Science, 22*(1), 8–15.
13. Ashby, M. F., Brechet, Y. J. M., Cebon, D., & Salvo, L. (2004). Selection strategies for materials and processes. *Materials and Design, 25*(1), 51–67.
14. Martin, J., Yahata, B., Hundley, J., et al. (2017). 3D printing of high-strength aluminium alloys. *Nature, 549*, 365–369.
15. Korpela, M., Riikonen, N., Piili, H., et al. (2020). Additive manufacturing—Past, present, and the future. In M. Collan & K. E. Michelsen (Eds.), *Technical, economic and societal effects of manufacturing 4.0*. Cham: Palgrave Macmillan.
16. Wiese, M., Thiede, S., & Herrmann, C. (2020). Rapid manufacturing of automotive polymer series parts: A systematic review of processes, materials and challenges. *Additive Manufacturing, 36*, 101582.
17. Johnson, N. S., Vulimiri, P. S., To AC, et al. (2020). Invited review: Machine learning for materials developments in metals additive manufacturing. *Additive Manufacturing, 36*, 101641.

Chapter 4
Role of Post-Process

1 How Process and Post-Process Are Related? What Are the Roles of Parameters in Optimization?

1.1 Stages of AM

AM process starts when materials and information are supplied to an AM system and shaping of the material starts. The process ends when shaping in the system ends. Before the shaping starts, the stage is a pre-process stage while after the shaping, it is a post-process stage (Fig. 4.1). Thus in-process stage (or process stage or main process stage) is preceded by a pre-process stage while followed by a post-process stage.

When material and information are fed into an AM system but the material is not started to interact with process parameters, the process is not yet started. For example, when laser powder bed fusion (LPBF) machine is filled up with powder, and digital file is transferred to the system but the powder layer is not yet laid on, the process is not started. Similarly, when a filament is fed to an extrusion based AM system but the filament is not yet started to melt, the process is not started. When the interaction between material and process parameters stops, the process stops.

1.2 Process and System

When the powder layer is laid on, the process is started. When the beam interacts with the powder layer, the process reaches to the next stage of the process. If the process stops at this point, one layer is solidified. It means some miniscule part of the product is formed. This is the evidence that the process has worked.

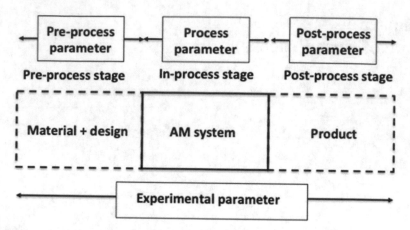

Fig. 4.1 Various stages of AM

When the powder layer is laid on and the process is started. If the process stops at this point, no layer is solidified. There is no any evidence left that the process has worked. The process does not leave a mark behind when the process is stopped before the beam interacts with the powder layer. The absence of a miniscule part or the absence of any evidence does not deny the process already starts when the powder layer is laid on. If the process fails to provide an evidence when it is stopped prematurely, it brings a question why this action of the process is considered important enough to be called the starting point of the process. Why not laying a powder layer should be ignored as the non-starting of the process.

When the powder layer is laid on, this happens because the process parameter sets in, the process parameter starts to interact with the material. This interaction changes powder stored in the container in the form of a layer. In the absence of parameters and interaction, there will not be any powder layer on a substrate or platform. It does not mean that the powder layer cannot be laid on a substrate in the absence of any parameter or interaction. It can be laid on, but this is not different from preparation of a machine or cleaning of a machine or warming a machine or cooling a machine or storing the file in the machine or checking the machine to verify its functions.

Starting the machine and starting the process are different. It does not mean that these actions of the machine are not important. These are important for a product to be formed. The formation of a product is also the objective of the process. A process requires parameters to set it on. In the absence of the parameters, the process does not start even if the arbitrary action of a machine will match the deliberate action of the process.

For example, AM machine or system starts in the anticipation of a product to be formed. During preparation of the machine, if a powder layer of an arbitrary thickness is laid on the platform and when the product needs to be formed and process parameters are required, the thickness of the layer matches with the layer thickness that is required to set the process in to form a product. In this example, the layer thickness arbitrarily set in the machine is equal to the layer thickness a process required for making a product. Thus, there is no need to lay the first powder layer again. Consequently, the process could not start by laying the layer. The process needs to go to the next stage, which is the interaction of the beam with the layer.

But, the arbitrary action of the machine cannot obscure the fact that the starting of the process is not without laying the layer. The arbitrary action of the machine cannot go far. Laying the layer is a recurring phenomenon. When the second layer will be laid on, the machine is helpless in the absence of process parameters. Consequently, it is the process that will decide the action of the machine.

It brings a question why it is important to know when a process starts. It is more important to know when it is known that the action of the process has no practical consequence when the process is stopped just after laying the first layer. The practical consequence means that the process is not strong enough to leave any slight evidence of its action if the process is stopped after laying the first layer.

In filament based AM, the process starts when the process parameters interact with the material, i.e., filament. This interaction leads to the melting of the filament and the forward movement of converted filament material. The converted filament material, which is henceforth an extrudate forms a layer. If the process is stopped at this point as the process was stopped earlier in the case of LPBF when the layer was laid, there is already a deposited and solidified layer. Though, the process is stopped and the next layer will not be deposited, but stopping the process after one layer does not fail the process to provide an evidence of its action. When the first layer is formed, some miniscule part of the product is already formed.

The process unlike LPBF shows the evidence of its action when the process is stopped after the first layer. It has proved that it does not do any futile action as LPBF does by laying the first layer. This is the first layer which creates the difference between two processes. This is the reason why it is important to know when a process starts.

Thus, there is a process which leaves a mark behind after the first layer and there is another process which does not leave. There is a process which does its duty of product fabrication form the first stage onwards and there is another process which does not do its duty from the first stage. Both processes are different but which one is better or which one is more important.

Whether action on the first stage of the process can be so much important to judge the importance of one process over the other. What will be the benefit of knowing the importance, whether it will have any industrial applications. Whether the so-called futile action will eventually lead to mass production (refer Chap. 8).

1.3 Post-Processing

When a process stops, the product forms. Stopping the process only implies that the shaping is stopped. The product is the result of the stopping of the shaping. Stopping the process and therefore stopping the shaping does not imply that the processed and shaped material (or the product) will no longer be processed and shaped. It only implies that further processing and shaping will no longer come under the realm of an AM process. If the product that is formed after the process stops is not the product that is expected, some more action is required to fulfill the expectation. This action is called post-processing.

1.4 Post-Processing Type Action During In-Process Stage

Milling, cutting, rolling, heating, etc. are some of the types of post-processing performed on an AM part. What if these post-processing will be used during in-process stage (Fig. 4.2).

During the in-process stage, while layer addition gives the shape of a part, some other actions can be used to improve the efficacy of the process. For example, in wire arc additive manufacturing (WAAM), after each layer or some layers, interpass rolling is used which improves the grain refinement [1]. In this case, rolling is the part of the process and is not a post-process since rolling is done before the completion of the shaping due to the process (Fig. 4.3). If rolling is done after the fabrication of the part, it will be a post-process (Figs. 4.5, 4.6) [2, 3].

In another example, after each layer or some layers, milling is used to fine finish the layer to improve the surface roughness [4]. In this case, milling is part of the process because milling is done before the fabrication of the part is completed. But if machining is done after the fabrication of the part, the machining comes under post-processing [5, 6].

The use of interpass rolling does not make WAAM [7] a non-AM process since it is WAAM that creates the shape by addition while interpass rolling does not create the shape. Though interpass rolling may decrease the height of the build and can influence the shape if the build has some fine features, but influencing the shape is not equal to creating the shape. Influencing the shape by rolling is a negative

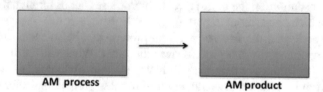

AM process   **AM product**

Fig. 4.2 AM product due to the presence of only AM process

Fig. 4.3 AM product because CM does not contribute to shaping

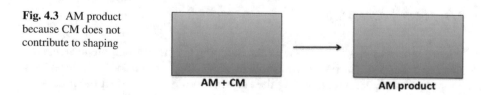

AM + CM   **AM product**

development which can be negated by bringing a change in a digital file—increasing or decreasing the size of various features of the file so that after the processing the part exactly represents the design.

When post-processing type action (such as machining) is part of a strategy to change the shape of the build so that a new shape is created that has contributions from both (AM and machining), then influencing the shape by machining is no longer a negative development but this is a necessity to make the shape. This development is no longer needed to be negated by the change in a digital file since this development is preplanned to create the shape that is not possible by AM alone.

The product thus formed is not shaped by one process (either AM or machining) alone at any stage of its development. The process responsible for creating this product is neither AM nor machining. But the process is a new process which is a mixture of AM and machining. The product thus formed is not an AM product (Fig. 4.4).

What if a product that is made from AM is gone through machining afterwards so that a new product having new shape is formed that is the result of AM and machining (Fig. 4.6b). The new product is not an AM product since the formation of its shape is due to both AM and machining. Though, this new non-AM product does not change the status of the process. Seeing the history of this non-AM product, it is clear at some stage of its development it was an AM product (Fig. 4.2).

Thus the same non-AM product can be the result of a mixture of processes (an AM process plus other process) or an AM process followed by other process (Figs. 4.4, 4.6). Having a non-AM product does not imply the absence of an AM process (Fig. 4.6) while having an AM product does not imply the need of absence of a post-process (Figs. 4.5, 4.6b).

When an AM process is followed by a post-processing to change the shape of the AM product, the resulting non-AM product is also known as four dimensional printing product [8].

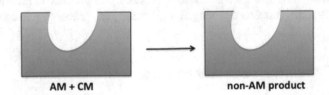

AM + CM   **non-AM product**

Fig. 4.4 Non-AM product because CM contributes to shaping

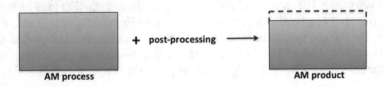

AM process   **AM product**

Fig. 4.5 AM product because post-processing does not create new shape

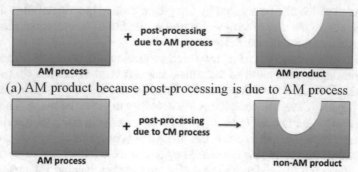

(a) AM product because post-processing is due to AM process

(b) Non-AM product because post-processing is due to CM

Fig. 4.6 AM and non-AM products when post-processing creates new shape. (**a**) AM product because post-processing is due to AM process, (**b**) non-AM product because post-processing is due to CM process

There can be five categories in which the generalized relation between AM process and post-process can be summarized. These categories are as follows:

- AM product due to AM
- AM product due to AM and CM
- Non-AM product due to AM and CM
- AM product due to AM and post-processing
- AM and non-AM products due to AM and post-processing

1.5   AM Product Due to AM

When an AM process is used to make a product, the product is an AM product (Fig. 4.2), its status remains unchanged until some post-processing is done.

1.6   AM Product Due to AM and CM

Before a product is formed, if AM process requires help of CM process to improve its performance but does not require help to create the shape, the resulting product is an AM product. For example, the use of intermediate machining [4, 9] when AM process is solely responsible for making the shape while machining is used to ensure that whatever shape is created by AM is not deviated from the original plan. In the absence of CM, AM will make a product but that product will be unsatisfactory. Thus the absence of CM will not prevent the product to acquire a shape.

1.7   Non-AM Product Due to AM and CM

The product is made due to contributions from both AM and CM. Before the product is formed, both AM and CM processes are required to contribute to the shape to the product [10]. In the absence of either AM or CM, even an unsatisfactory product will not be formed. Since the product formed is not solely due to AM, the product is a non-AM product.

1.8   AM Product Due to AM and Post-Processing

AM process is used to make a product and the product is consequently an AM product. But the product requires post-processing to improve properties, post-processing can change the size of the product but the shape of the product is solely due to AM process [2, 3]. The post-processing does not contribute to the shape, and therefore post-processing is unable to change the status of an AM product.

1.9   AM and Non-AM Products
Due to AM and Post-Processing

If an AM product requires further action to change its shape, the post-processing is used for the product to have a final shape. The product thus formed is not only due to AM but also due to CM, the product is therefore a non-AM product (Fig. 4.6b).

But if another AM process is used to perform post-processing, there is no presence of a non-AM process to make it a non-AM product, and the product is an AM product (Fig. 4.6a) [11]. Connecting both process and post-process digitally will lead to digital manufacturing (refer Chap. 9).

1.10   Experimental and Process Parameters

All parameters that are used to do experiments are experimental parameters. They can be pre-process, process, and post-process parameters (Fig. 4.1). For example, if a steel product is to be made using LPBF, parameters can be type of powders, laser power, scan speed, heat treatment. Out of these parameters, type of powders is a pre-process parameter while laser power and scan speed are process parameters, and heat treatment is a post-process parameter [12]. Thus any parameters such as pre-process parameters, process parameters, and post-process parameters can be experimental parameters, but all experimental parameters may not be process parameters.

1.11 Machine and Process Parameters

The process parameters are machine parameters as the process requires a machine to be executed. In the above example, laser power and scan speed are process parameters requiring LPBF machine (to execute the process LPBF), and are machine parameters as well. Another parameter, i.e., type of powders is a pre-process parameter and is not available in LPBF machine and this parameter cannot be changed in machine and, consequently, is not a machine parameter. Another parameter, i.e., heat treatment is a post-process parameter and requires another machine, i.e., furnace to be executed and thus is not a (LPBF) machine parameter.

What if pre-process and post-process parameters can be executed in the same machine. Imagining a state when LPBF machine is developed that many types of powders can be stored and changed as per need—after making a steel product using one type of powders, the machine automatically gets emptied and is filled with another type of powders to make another steel product. Though parameter, i.e., type of powders is a pre-process parameter but in the presence of facility to change it within LPBF machine, it is a machine parameter.

Thus depending on the provision within a machine; pre-process, process, and post-process parameters can be machine parameters. And if experiment is done using all available parameters in a machine, experimental parameters are machine parameters.

What if heat treatment is done inside LPBF machine, which is possible with the help of existing lamp or heater or laser beam. When the part is fabricated inside LPBF machine and instead of taking it out to a furnace for heat treatment, the part is left as it is inside the machine, the heat treatment performed on the part is a post-process action. Therefore, it is a post-process parameter. Since it is done inside the machine, it is a machine parameter as well. But, it is still different from other parameters (laser power, scan speed) that are executed in the same machine. In the case of other parameters, the boundary of the process and the machine is same; they are both process and machine parameter. But, heat treatment is not a process parameter. Though, it can become a process parameter in other examples.

If it is done after each layer or a couple of layers, it is not a post-process but process parameter. It is done for decreasing residual stress [13] or surface roughness or increasing accuracy. Though heat treatment is itself a process on its own right, but its incorporation into another process, i.e., LPBF does not dilute LPBF for being an AM process. This reason is not different from why incorporation of interpass rolling still lets WAAM to be an AM process.

1.12 Optimization

Optimization of parameter means finding the right value of parameters that gives the best product. If application changes, the definition of a best product changes and then the right value of changes. For example, if a polymer bearing is made using LPBF, it must have high surface finish. In this example, a product is best when it has high surface finish. If a door handle is made from the same polymer using the same process, the requirement is not to have a high but a low surface finish so that the handle can provide adequate grip in use. In this case, the product is best when it has a low surface finish, and the definition of a best product in this case is not the same as that was in a previous case. Thus if the aim is to have high surface finish, it will lead to one set of optimized parameters while the aim of the low surface finish will lead to another.

This brings a question which parameters need to be optimized. In most AM machines, machine parameters are process parameters. It is convenient to optimize machine parameters to find a product. But the best product obtained may not be the best if seen from a larger perspective (pre-process plus process plus post-process). To find the best product, all parameters from the gamut of whole process chain need to be selected. However, for understanding the process it is the optimization of process parameters rather than all parameters are required.

If optimization of process parameters is done with an aim to find the best product for a given application, it will inform whether a particular process for a particular application is best or not. Optimization of all parameters (total optimization) instead of process parameters may furnish better products than that are possible only with the optimization of process parameters, but the advantage thus gained from total optimization may not let uncover the deficiency of a process. Consequently, the absence of understanding a process will not help controlling the process.

2 What Is the Role of Post-Process When AM Is Not Efficient? Is Cleaning a Post-Process?

2.1 Role of Post-Process

An efficient AM process provides required properties such as strength, density, dimension, etc. while an inefficient AM process is not able to provide these properties even if there is a requirement.

Post-process is used when a product made in the in-process stage is not final. Thus the post-process is used as a complementary process to make up the demerit of the inefficient main process, or to fulfil the requirement of a product which cannot be achieved in the in-process stage. Following are the examples of complementary post-processing.

A metallic part made from LPBF is having residual stress because of high rates of beam–material interaction [14]. Post-process as heat treatment is used to remove the stress [15]. In this case, LPBF is the main AM process while heat treatment is a post-process. LPBF cannot make a part (residual-) stress-free. It is the demerit of LPBF. Therefore, post-process is required to make up the demerit. If LPBF is efficient enough to make a stress-free part, possibly using in-situ layer treatment [13] or island scan strategy [16], there is no need for a post-process.

In another example, a polymer model for a mobile phone cover is made using photopolymer bed process (PPBP) and is painted by applying coating. In this case, PPBP is a process while coating is a post-process. Since PPBP does not make a colorful part in many colors, a post-process is required to make a colorful part in a desired color [17]. Thus coating as a post-process is used to compensate for the demerit of main process (PPBP). Though this post-process is only for the purpose of decoration and does not change the design, dimension, and mechanical properties of a part, or add or remove any features of a part but is still valuable because AM is not efficient to do it on its own.

2.2   Cleaning

Following examples will clarify whether cleaning is significant. Two products are made from electron beam powder bed fusion (EPBF): one is a knee implant, and the other is a nozzle for laser cladding machine. Since knee implant needs to be cleaned before it is fitted inside a patient, the cleaning is significant. EPBF cannot make a knee implant that is, without being cleaned, directly fitted inside the patient. Thus there is a demerit in EPBF for not able to do both—making a part and subsequently cleaning it. The demerit requires to be overcome with the help of post-process, i.e., cleaning. This makes it significant.

In the case of a nozzle made by the same process (EPBF), lack of cleaning does not deprive the nozzle from being fitted in a laser cladding machine and from getting used subsequently. The efficiency of the nozzle does not change irrespective of it being cleaned or not. Thus EPBF is an efficient process in this case and does not require any post-process (cleaning) as a necessity. In this case, there is no demerit in EPBF that needs to be overcome with the help of a post-process. If cleaning is used, it is not a significant action to be a post-process.

But in the same nozzle, if there are some powders metallurgically bonded on its surface, these powders need to be removed first to allow the nozzle to be used efficiently. The cleaning for removing powders is not insignificant and is a post-process. Other examples are cleaning for increasing the accuracy of parts after manufacturing [18] or after support removal [19].

Thus it overcomes the demerit of the process in the case of a knee implant and an inaccurate nozzle and is a post-process. While in the case of a nozzle having no powders attached on its surface, it does not have opportunity to overcome any demerit (as there are no demerits) of the main process and is not a post-process.

3 Which One Should Be the Priority—Process or Process Chain?

3.1 Example

Considering an example of LPBF when a composite needs to be manufactured. The composite is due to two materials: an iron-based alloy and a copper-based alloy. A part from an iron-based alloy needs to be formed using LPBF, and is infiltrated thereafter with a copper-based alloy so that the final result will be a composite consisting of both alloys. In this example, the main process is LPBF while the post-process is infiltration. This method is used to make plastic injection molds [20, 21].

A dense part is not amenable to infiltration (Fig. 4.7a). Besides, closed pores do not allow infiltrant to move and reach deep inside the part (Fig. 4.7b). If a part is porous, it allows infiltrant to infuse it during post-processing (Fig. 4.7c), which results in a high strength composite. Figure 4.7 shows infiltration of parts (black in color) having pores (white in color) with infiltrants (red in color). The result of the infiltration is shown by a change in color from white to red. For a dense part, infiltrants cannot enter inside the part and are remained outside at the boundary; thus the color of only the boundary is changed (Fig. 4.7a).

3.2 Inference

Making a porous part shows that LPBF is not used to its maximum efficiency to make the part. If the process was used to its maximum efficiency to make a fully dense part, there would be no scope for effective infiltration (Fig. 4.1a) and then there would be no desired composite.

When the process is not used to make a fully dense part, the fabrication of a pre-part leads finally to the fabrication of the desired part. Thus, the process is using only a part of its efficiency—nevertheless, it does not stop the desired part to be fabricated in the long run. It is because the process is combined with a post-process. Though the main process and its efficiency are important but what is more important is a strategy—a strategy which combines process and post-process [22] for product development [23], a strategy which underutilizes a process because a product is more important than a process, or the product after a process chain (process plus post-process) is more important than that after a main process. The strategy shows that an efficient process is not efficient if it does not lead to a desired product at the end (refer Chap. 7).

The strategy brings questions—if the process chain is more important than a process, why an inefficient process should be converted to an efficient one. Why not all efforts should be made to search for a strategy, and an inefficient process should be left as it is because there requires an intensive effort to find a matching post-process, or strategy should be emphasized in a way that all focus will shift towards developing a process chain rather than a process.

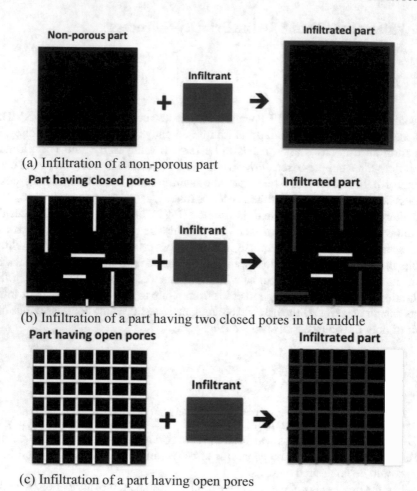

(a) Infiltration of a non-porous part

(b) Infiltration of a part having two closed pores in the middle

(c) Infiltration of a part having open pores

Fig. 4.7 Infiltration of (**a**) a non-porous part, (**b**) a part having closed pores, (**c**) a part having open pores

The strategy does not undermine main AM process development but informs that process development should be judged from a product development view. Whether a process is developed or not will be known from whether the process leads to a desired product. Developing a process does not always work and therefore in many cases it is the process chain that works rather than developing a process [24]. Besides, an inefficient process is no longer inefficient if it is combined with a post-process. The strategy does not discourage development of a main AM process but apprises that an undeveloped process is not a failure in all cases and can be game-changer. The main process must be developed but there should not be a wait because even before the development, there can be some strategy that will lead to a success-ful product development with the help of an undeveloped process [25].

4 What Are the Types of Post-Processing Used in AM?

The post-processing used in AM are listed as follows:

- Barrel finishing [26],
- Chemical polishing: reduces surface roughness to sub-micron level [27],
- Chemical vapor infiltration [28],
- Coating [29],
- Cold isostatic pressing [30],
- Curing: increases mechanical properties but changes the dimension [31, 32],
- Cutting [5],
- Drilling [33],
- Electrochemical polishing: requires space for electrodes, convenient for simple parts [27],
- Electrochemical infiltration [34],
- Electroplating [35],
- Etching: time-taking process, suitable for big parts [36],
- Fine finishing [37],
- Forging [38],
- Grinding: suitable for non-complex parts [39],
- Heating: for microstructure improvement [40, 41],
- Hot isostatic pressing: for removing small pores [42, 43],
- Infiltration [44],
- Laser peening or laser shock peening: for improving fatigue properties [45, 46],
- Machining [47],
- Milling [48],
- Painting [49],
- Rolling [2, 3],
- Shape changing with the help of environment (4D printing) [50],
- Shot-peening: for improving fatigue properties [42, 51],
- Stress-relieving [14],
- Sintering [52],
- Support removal [53],
- Welding [54].

References

1. Derekar, K. S. (2018). A review of wire arc additive manufacturing and advances in wire arc additive manufacturing of aluminium. *Materials Science and Technology, 34*(8), 895–916.
2. Zhao, Z., Tariq, N. H., Tang, J., et al. (2020). Microstructural evolutions and mechanical characteristics of Ti/steel clad plates fabricated through cold spray additive manufacturing followed by hot-rolling and annealing. *Materials and Design, 185*, 108249.

3. Sokolov, P., Aleshchenko, A., Koshmin, A., et al. (2020). Effect of hot rolling on structure and mechanical properties of Ti-6Al-4V alloy parts produced by direct laser deposition. *International Journal of Advanced Manufacturing Technology, 107*, 1595–1603.

4. Karunakaran, K. P., Suryakumar, S., Pushpa, V., & Akula, S. (2010). Low cost integration of additive and subtractive processes for hybrid layered manufacturing. *Robotics and Computer-Integrated Manufacturing, 26*(5), 490–499.

5. Oyelola, O., Crawforth, P., M'Saoubi, R., & Clare, A. T. (2018). On the machinability of directed energy deposited Ti6Al4V. *Additive Manufacturing, 19*, 39–50.

6. Yamazaki, T. (2016). Development of a hybrid multi-tasking machine tool: Integration of additive manufacturing technology with CNC machining. *Procedia CIRP, 42*, 81–86.

7. Wu, B., Pan, Z., Ding, D., et al. (2018). A review of the wire arc additive manufacturing of metals: Properties, defects and quality improvement. *Journal of Manufacturing Processes, 35*, 127–139.

8. Momeni, F., & Ni, J. (2020). Laws of 4D printing. *Engineering, 6*(9), 1035–1055.

9. Du, W., Bai, Q., & Zhang, B. (2016). A novel method for additive/subtractive hybrid manufacturing of metallic parts. *Procedia Manufacturing, 5*, 1018–1030.

10. Chen, N., & Frank, M. (2019). Process planning for hybrid additive and subtractive manufacturing to integrate machining and directed energy deposition. *Procedia Manufacturing, 34*, 205–213.

11. Petrat, T., Graf, B., Gumenyuk, A., & Rethmeier, M. (2016). Laser metal deposition as repair technology for a gas turbine burner made of inconel 718. *Physics Procedia, 83*, 761–768.

12. Dowling, L., Kennedy, J., O'Shaughnessy, S., & Trimble, D. (2020). A review of critical repeatability and reproducibility issues in powder bed fusion. *Materials and Design, 186*, 108346.

13. Roehling, J. D., Smith, W. L., Roehling, T. T., et al. (2019). Reducing residual stress by selective large-area diode surface heating during laser powder bed fusion additive manufacturing. *Additive Manufacturing, 28*, 228–235.

14. Li, C., Liu, Z. Y., Fang, X. Y., & Guo, Y. B. (2018). Residual stress in metal additive manufacturing. *Procedia CIRP, 71*, 348–353.

15. Tomus, D., Tian, Y., Rometsch, P. A., et al. (2016). Influence of post heat treatments on anisotropy of mechanical behaviour and microstructure of Hastelloy-X parts produced by selective laser melting. *Materials Science and Engineering A, 667*, 42–53.

16. Masoomi, M., Thompson, S. M., & Shamsaei, N. (2017). Laser powder bed fusion of Ti-6Al-4V parts: Thermal modeling and mechanical implications. *International Journal of Machine Tools and Manufacture, 118–119*, 73–90.

17. Gersch, M., Tuncer, I., Hotter, A., & Pfefferkorn, F. (2017). Methods for coating objects in particular such objects that have been manufactured by a generative manufacturing method. US Patent 9586369.

18. Schwarzer, E., Götz, M., & Markova, D. (2017). Lithography-based ceramic manufacturing (LCM) – Viscosity and cleaning as two quality influencing steps in the process chain of printing green parts. *Journal of the European Ceramic Society, 37*(16), 5329–5338.

19. Verhaagen, B., Zanderink, T., & Rivas, D. F. (2016). Ultrasonic cleaning of 3D printed objects and cleaning challenge devices. *Applied Acoustics, 103B*, 172–181.

20. Kumar, S. (2008). Microstructure and wear of SLM materials. In: *SFF Proceedings* (pp. 128–142).

21. Kumar, S. (2009). Sliding wear behaviour of dedicated iron-based SLS materials. *International Journal of Advanced Manufacturing Technology, 43*(3), 337–347.

22. Häfele, T., Schneberger, J. H., Kaspar, J., et al. (2019). Hybrid additive manufacturing – Process chain correlations and impacts. *Procedia CIRP, 84*, 328–334.

23. Kaspar, J., Bechtel, S., Häfele, T., et al. (2019). Integrated additive product development for multi-material parts. *Procedia Manufacturing, 33*, 3–10.

24. Biondani, F. G., Bissacco, G., Mohanty, S., et al. (2020). Multi-metal additive manufacturing process chain for optical quality mold generation. *Journal of Materials Processing Technology, 277*, 116451.

25. Kumar, S. (2018). Process chain development for additive manufacturing of cemented carbides. *Journal of Manufacturing Processes, 34*, 121–130.
26. Boschetto, A., Bottini, L., Macera, L., & Veniali, F. (2020). Post-processing of complex SLM parts by barrel finishing. *Applied Sciences, 10*, 1382.
27. Tyagi, P., Goulet, T., Riso, C., et al. (2019). Reducing the roughness of internal surface of an additive manufacturing produced 316 steel component by chempolishing and electropolishing. *Additive Manufacturing, 25*, 32–38.
28. Zhu, Q., Dong, X., Hu, J., et al. (2020). High strength aligned SiC nanowire reinforced SiC porous ceramics fabricated by 3D printing and chemical vapor infiltration. *Ceramics International, 46*(5), 6978–6983.
29. Yao, S., & Wang, T. (2016). Improved surface of additive manufactured products by coating. *Journal of Manufacturing Processes, 24*(1), 212–216.
30. Liu, K., Sun, H., Shi, Y., et al. (2016). Research on selective laser sintering of kaolin–epoxy resin ceramic powders combined with cold isostatic pressing and sintering. *Ceramics International, 42*(9), 10711–10718.
31. Zhao, J., Yang, Y., & Li, L. (2020). A comprehensive evaluation for different post-curing methods used in stereolithography additive manufacturing. *Journal of Manufacturing Processes, 56A*, 867–877.
32. Jindal, P., Juneja, M., Bajaj, D., et al. (2020). Effects of post-curing conditions on mechanical properties of 3D printed clear dental aligners. *Rapid Prototyping Journal, 26*, 1337.
33. Rysava, Z., Bruschi, S., Carmignato, S., et al. (2016). Micro-drilling and threading of the Ti6Al4V titanium alloy produced through additive manufacturing. *Procedia CIRP, 46*, 583–586.
34. Goel, A., & Bourell, D. (2011). Electrochemical deposition of metal ions in porous laser sintered inter-metallic and ceramic preforms. *Rapid Prototyping Journal, 17*(3), 181–186.
35. Yang, G., Yu, S., Mo, J., et al. (2018). Bipolar plate development with additive manufacturing and protective coating for durable and high-efficiency hydrogen production. *Journal of Power Sources, 396*, 590–598.
36. Raikar, S., Heilig, M., Mamidanna, A., & Hildreth, O. J. (2020). Self-terminating etching process for automated support removal and surface finishing of additively manufactured Ti-6Al-4V. *Additive Manufacturing, 37*, 101694.
37. Yamaguchi, H., Fergani, O., & Wu, P. (2017). Modification using magnetic field- assisted finishing of the surface roughness and residual stress of additively manufactured components. *CIRP Annals - Manufacturing Technology, 66*(1), 305–308.
38. Jiang, J., Hooper, P., Li, N., et al. (2017). An integrated method for net-shape manufacturing components combining 3D additive manufacturing and compressive forming processes. *Procedia Engineering, 207*, 1182–1187.
39. Lober, L., Flache, C., Petters, R., et al. (2013). Comparison of different post processing technologies for SLM generated 316l steel parts. *Rapid Prototyping Journal, 19*(3), 173–179.
40. Maamoun, A. H., Elbestawi, M., Dosbaeva, G. K., & Veldhuis, S. C. (2018). Thermal post-processing of AlSi10Mg parts produced by selective laser melting using recycled powder. *Additive Manufacturing, 21*, 234–247.
41. Zhang, M., Yang, Y., Song, C., et al. (2018). Effect of the heat treatment on corrosion and mechanical properties of CoCrMo alloys manufactured by selective laser melting. *Rapid Prototyping Journal, 24*(7), 1235–1244.
42. Qian, M., Xu, W., Brandt, M., & Tang, H. (2016). Additive manufacturing and postprocessing of Ti-6Al-4V for superior mechanical properties. *MRS Bulletin, 41*(10), 775–784.
43. Kreitcberg, A., Brailovski, V., & Turenne, S. (2017). Effect of heat treatment and hot isostatic pressing on the microstructure and mechanical properties of inconel 625 alloy processed by laser powder bed fusion. *Materials Science and Engineering, 689*, 1–10.
44. Li, L., Tirado, A., Conner, B. S., et al. (2017). A novel method combining additive manufacturing and alloy infiltration for NdFeB bonded magnet fabrication. *Journal of Magnetism and Magnetic Materials, 438*, 163–167.

45. Hackel, L., Rankin, J. R., Rubenchik, A., et al. (2018). Laser peening: A tool for additive manufacturing post-processing. *Additive Manufacturing, 24*, 67–75.
46. Kalentics, N., Boillat, E., Peyre, P., et al. (2017). Tailoring residual stress profile of selective laser melted parts by laser shock peening. *Additive Manufacturing, 16*, 90–97.
47. Stucker, B., & Qu, X. (2003). A finish machining strategy for rapid manufactured parts and tools. *Rapid Prototyping Journal, 9*(4), 194–200.
48. Lopes, J. G., Machado, C. M., Duarte, V. R., et al. (2020). Effect of milling parameters on HSLA steel parts produced by wire and arc additive manufacturing (WAAM). *Journal of Manufacturing Processes, 59*, 739–749.
49. Nsengimana, J., Van der Walt, J., Pei, E., & Miah, M. (2019). Effect of post-processing on the dimensional accuracy of small plastic additive manufactured parts. *Rapid Prototyping Journal, 25*(1), 1–12.
50. Momeni, F., Hassani, S. M. M., Liu, X., & Ni, J. (2017). A review of 4D printing. *Materials and Design, 122*, 42–79.
51. AlMangour, B., & Yang, J. M. (2016). Improving the surface quality and mechanical properties by shot-peening of 17-4 stainless steel fabricated by additive manufacturing. *Materials and Design, 110*, 914–924.
52. Sheydaeian, E., Sarikhani, K., Chen, P., & Toyserkani, E. (2017). Material process development for the fabrication of heterogeneous titanium structures with selective pore morphology by a hybrid additive manufacturing process. *Materials and Design, 135*, 142–150.
53. Nelaturi, S., Behandish, M., Mirzendehdel, A. M., & Kleer, J. (2019). Automatic support removal for additive manufacturing post processing. *Computer-Aided Design, 115*, 135–146.
54. Wits, W. W., & Becker, J. M. J. (2015). Laser beam welding of titanium additive manufactured parts. *Procedia CIRP, 28*, 70–75.

Chapter 5
Comparison

1 Why Is Complexity Different in AM Than in Machining?

In additive layer manufacturing, 2D layers are formed which on aggregation give a 3D part. When a complex part needs to be formed, there is a need to aggregate complex 2D layers. Since the complexity of any 2D layer of a part does not exceed the complexity of a part, the fabrication of a difficult complex part becomes the fabrication of non-difficult complex 2D layers. Thus, higher difficulty of the fabrication of a 3D part becomes lower difficulty of the fabrication of 2D layers. AM does not visualize a complex part as complex but as an aggregation of lesser complex 2D layers. Due to this aggregation, there exists problem less about complex 3D part and more about less complex 2D layers.

Perception of complexity differently in AM has practical implications as well. Increasing complexity does not increase difficulty in manufacturing, cost of manufacturing [1], and manufacturing time in the same way as it increases in machining. If there are two designs, one is simple and the other is complex, it is straightforward to say which one is complex to fabricate in machining—simple design means simple fabrication while complex design means complex fabrication. The design is, in most cases, the basis for determining the complexity in fabrication. But in AM, if the complexity between two designs is not very different, it is difficult to determine which one is more complex.

For example, if two designs have each a combination of different numbers of features such as thin walls, pins, cavities, overhangs, undercuts, it is difficult to identify a design of higher complexity. If it is quantified that one of two designs is more complex, it does not give guarantee that which one is more complex to fabricate in AM. Even if it is known which one is more complex to fabricate using a particular AM process, it is again not certain which one is more complex to fabricate using a particular AM machine. As benchmark studies have shown that a given

S. Kumar, *Additive Manufacturing Solutions*,
https://doi.org/10.1007/978-3-030-80783-2_5

design leads to different degrees of successes in fabrication using a particular AM process but different AM machines [2].

2 Why Design for AM Is Different from Design for Machining?

There are two reasons why design for AM is different from design for machining. Firstly, there are many tools of various sizes in machining, and secondly change in design orientation during fabrication is possible in machining (refer Chap. 2).

2.1 *Machining Has Many Tools*

There are many tools in machining to choose from while AM does not have such number of tools. If a design consists of both big and small features, these will be made with big and small tools, respectively, in machining. For example, if a design consists of a thin and a thick wall, they will be machined out from a block using a small and a big milling tool, respectively (Fig. 5.1a). Thus, machining has an advantage to make various features by selecting tools of matching sizes. Such freedom to choose from many tools does not exist in AM.

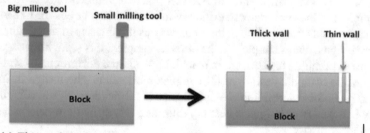

(a) Thin and thick walls are machined by small and big milling tools respectively (side view)

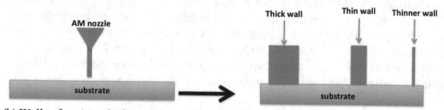

(b) Walls of various thickness are made by a single nozzle (side view)

Fig. 5.1 Walls made by (**a**) milling and (**b**) AM

In AM, it is not difficult to find an equivalent object that can be considered the counterpart of a machining tool. For example, in a beam based AM, this object can be beam. It can be laser in laser powder bed fusion (LPBF), electron beam in electron beam powder bed fusion (EPBF), nozzle in material jetting or extrusion, etc.

This single object or tool in AM has to do all the jobs that are usually done by many tools of various sizes in conventional manufacturing (CM) such as drilling or milling. In LPBF, it is the same laser beam that is used to make both big and small walls or both wide and narrow cavities. In deposition process (DP), it is the same nozzle which is used to extrude or jet material for walls (Fig. 5.1b) and cavities of all sizes.

If a design is to be made for AM, it has to be taken into account that various features have to be made by a single tool (Fig. 5.1b) that does not give the same accuracy and surface finish as when the same features are made in machining. If a design is to be made for machining, accuracy of a feature depends more on the selection of the right tool than on the optimization of parameters, as having option to be able to select the right tool takes some burden away from optimizing the parameters.

2.2   Making Wall Using Multi-Tool AM

What if AM tries to compete with machining by using a number of tools (beams or nozzles). Considering AM uses two types of beams—one of a big spot size to make big features and the other of a small spot size to make small features; or two types of nozzles—one of a big outlet diameter to make big features and the other of a small outlet diameter to make small features.

Using two types of beams or nozzles will provide advantages in making small features which cannot be made when AM has only one beam or nozzle of bigger dimension. For example, if a thin wall of 400 μm thickness is to be fabricated in LPBF or EPBF, a beam of 100 μm rather than 500 μm diameter will be more suitable. Using a small diameter will provide a small melt pool and then a small inaccuracy between two consecutive melt pools that will help miniaturize the process and make small features. Thus, it is possible to make all thick walls using a smaller diameter beam, though the properties of the walls will not be the same [3].

If a thick wall of 20 mm thickness is to be fabricated, a beam of 500 μm rather than 100 μm diameter will be more suitable. Since 500 μm diameter will speed up the process while 100 μm diameter will take longer time and slow down the process. But, 100 μm diameter will still be able to make the thick wall and since 100 μm diameter gives a higher accuracy, both side and main surfaces will have higher accuracies. Thus for speeding up the process without having to make small features, a big diameter beam is suitable while for the accurate fabrication of a complex product, a small diameter beam is suitable. However, the speed advantage achieved with a big diameter beam is limited, the layer thickness or scan speed has more influence on speed than the diameter has [4]. Therefore, to speed up the fabrication using a big

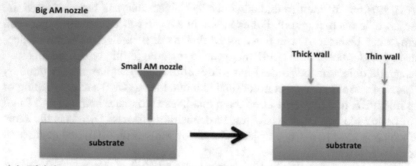

(a) Thick and thin walls made by big and small nozzle respectively (side view)

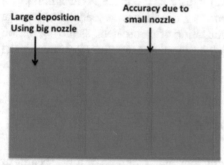

(b) Structureconsisting of thick and thin walls (top view)

Fig. 5.2 Deposition due to multi-nozzle DP: (**a**) thick and thin walls made by big and small nozzle, respectively (side view), (**b**) structure consisting of thick and thin walls (top view)

diameter beam in AM is not a strategy as right as using a big milling tool in machining.

If a big diameter nozzle is used, it will extrude a large amount of material (Fig. 5.2a) in a given time and will accelerate the fabrication while a small diameter nozzle used along with can improve the accuracy of a part (Fig. 5.2b).

If two beams of different diameters are used in AM, it will:

- increase cost,
- increase complexity in algorithm for processing a layer,
- require hardware adjustment in a single machine,
- require planning to determine the sequence of processing various features using one of beam diameters,
- require decision to find which intermediate size feature will come under small or big features,
- require to determine overlap at the boundary between two features processed by two types of beams. Thus single beam instead of two beams is more convenient to be used.

Using many beams or spots increases the productivity [5], but in the absence of right adjustment between scanning by two beams, the quality deteriorates. If two beams are used, the plume (gas plus small particles) generated by one beam attenuates other beam. Hence, the part made using two beams results in lower density [6]. At extreme cases, a big diameter beam gives residual stress while a small diameter beam vaporizes the powder; therefore, there is a limitation to the maximum difference in diameters that can be used.

2.3   Making Cavity Using Multi-Tool AM

In machining, if a drilling tool of diameter 100 μm is to process a design consisting of cavities of diameters 100 μm, 500 μm, 2 mm, and 2 cm, this single tool will not be able to make cavities of any diameter other than 100 μm. It is because the tool is bound to make a cavity having internal diameter similar to only its diameter by removing materials from an area equal to its diameter. While in AM, a 100 μm diameter beam can make cavities of other diameters as well.

In AM, a beam does not dig out a cavity as a drilling tool does (Fig. 5.3a) but processes adjacent areas (Fig. 5.3b) so that a circular wall for a cavity is made. When the wall is deposited, the area enclosed within the wall gets the shape of the cavity—higher wall leads to a deep cavity while the bigger circumference of the wall leads to a wide cavity.

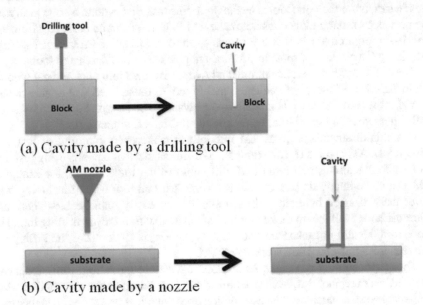

(a) Cavity made by a drilling tool

(b) Cavity made by a nozzle

Fig. 5.3 Cavity made in drilling (**a**) and AM (**b**)

How small a cavity can be—as long as the solidification at diametrically opposite points at the cavity does not overlap so that the desired space within the cavity does not shrink. If the size of the cavity shrinks, the size of the solidified material needs to be made smaller [7]. This can be made smaller if feedstock size such as powder or wire in the case of solid feedstock based processes is made smaller, or photopolymer liquid of low curing capacity in the case of photopolymer based process is used [8], or beam diameter size is made smaller so that the effect of the beam is restricted to a smaller area on a layer.

2.4   Inference

Machining is better than AM if AM is not developed. Since, in machining, there is a guarantee that either a design will be made or not be made. But in AM, there is mostly one tool that has to make all features and therefore if AM is not well developed, a complex design will not be properly converted to a part.

In machining, there are tools of standard sizes available to make various sizes of features; therefore, a design which requires only those tools will be converted to a part. The part will have better surface finish and accuracy. In AM, in the absence of any feature-specific tool, one tool needs to make all features. Consequently, AM is expected to be developed enough so that its one tool will make almost everything required in a design. If one tool is making all sizes of features, the tool is not expected to perform better than a bunch of dedicated tools.

AM gets better in some cases when it is equipped with more than one tool. In these cases, more than one tool is available for the task that is done by one tool. But, the presence of many tools does not make AM tool free from its basic requirement. The basic requirement is that a tool has to make features of various sizes. Thus increasing the number of tools in AM does not make a tool a dedicated tool.

The difference between machining and AM is that machining has dedicated tools while there is no concept of dedicated tool (a tool to make fixed size or size range) in AM. Absence of such tools in AM is a disadvantage because the tool comes along with a guarantee in fabrication, better surface finish, and dimension.

But, this disadvantage gives AM freedom to make everything out of just one thing (tool). AM can make, for example, 20 different sizes of cavities using just one tool while machining will require 20 different drilling tools. Again, for example, AM can make hundred sizes of cavities using just one tool while machining will never make it since hundred drilling tools are not easily manageable. There are other methods available to make hundred of cavities without using drilling tool, but these methods are not taken into account as the aim is to compare the difference between machining tool and working of AM.

Though machining can make 20 different cavities because 20 drilling tools can be fitted in a machine. But, what machining can do is based on the assumption what is available on the machine. The assumption does not take into account whether tool will reach to the point of interest, how process planning will be done, whether different sizes of tools will require same or different fixtures, etc. AM is free from such potential problems associated with such tools. This is the advantage of AM.

2.5   Difference Between AM and Machining

In AM, building happens only through top surface. It means only the top surface is engineered. Other surfaces such as side and bottom surface are the result of making the top surface. Thus, the main outcome of AM is the top surface while the bottom and side surfaces are the secondary outcomes. Bottom surface is the result of how it is detached from the substrate (if it is attached) or what is the condition of the platform during building (if it is not attached).

It does not mean if side surfaces are given attention, these cannot be improved. Improvement happens by the usual methods such as parameter optimization or orientation changing, but the improvement happens only by applying such methods on the top surface. The basic requirement of AM, that whenever top surface will be made the side surface will be automatically made, does not change. The presence of the basic requirement means there is no guarantee that top and side surface will have the same quality or roughness or planarity.

But, the basic requirement can be defeated by some means. The ongoing build can be stopped, the orientation of the block can be changed by 90° so that the top surface is no longer the top surface, the side surface is no longer the side surface. The side surface becomes henceforth the top surface. Since, the problem is with the side surface, when the side surface becomes the top surface, it can be engineered and controlled better. The problem can thus be minimized if geometry allows such processing. But the problem can only be minimized and cannot be completely solved because the action of creating a surface will always create a side surface; again changing the block by 90° (to catch and engineer the side surface) will not be practical.

Thus if AM is improved so that some DP changes the ongoing build [9], the difference between the side surface and the top surface can be minimized but making a new type of system to change the build direction will not make it free from the basic requirement. In AM, side surface is generally managed by having two parameters for processing a layer—one parameter for processing the boundary of the layer while other parameter is for processing the non-boundary area of the layer. Thus, the side surface is the outcome of how the boundary of the layer is processed.

In machining, surfaces are already got before the machining starts, the surfaces remain as they are until disturbed by the machining. There is no interconnection between the side and the top surface in machining as AM has.

For example, if milling is done on the top surface of a block to make a rectangular cavity, action through the top surface creates four side surfaces, i.e., four internal walls of the cavity. But the creation of these four side surfaces does not directly depend on the working of the top surface. Once a smaller cavity than the required cavity size is made by working on the top surface, the milling tool does not need to work on the top surface anymore to create the required cavity size.

The tool instead works on the side surface directly to make required bigger cavity. The surface quality of the side surface depends upon what the milling tool does on the side surface. The surface quality of the side surface does not depend upon what the milling tool earlier did on the top surface when the milling tool was making smaller cavity as a precursor to make bigger cavity. Thus, in machining, the side surface of the cavity is not the direct outcome of the processing of the top surface.

If the tool works on the whole top surface of the block instead of partially working on the top surface to make a cavity, the top layer of the block is machined away or removed. With the removal of the top layer, the side surface of the top layer is also gone (Fig. 5.4a). Thus if milling tool reaches the side surface of the block, the side surface is no longer present, and is removed. If the tool does not reach the side surface of the block, the side surface remains as it is. The side surface will remain as it is until disturbed by the tool.

This is the basic difference between AM and machining. In AM, while a top surface is processed (made), the side surface is already formed (Fig. 5.4b). In machining, while the top surface is processed (removed), the side surface is removed along with. This is the advantage of machining over AM. In AM, the side surface is already made which needs to be accepted as it is, if there is problem on the side

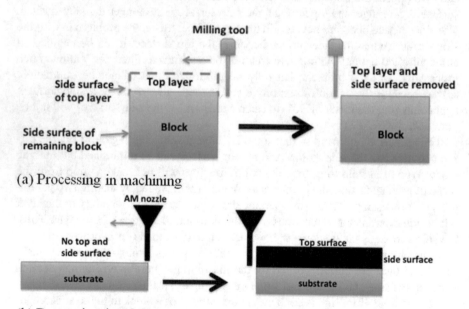

(a) Processing in machining

(b) Processing in AM

Fig. 5.4 Processing in machining (**a**) and AM (**b**)

surface, the problem cannot be removed by AM anymore (secondary process is required). In machining, the side surface is already gone, and the problem is gone along with the side surface. If in cavity, there are some problems on the side surface, the problem can be removed by machining again.

When the milling tool removes the top layer of a block, side surface associated with the top layer also goes. But, the side surface of the remaining portion of the block (Fig. 5.4a) can get damaged by the tool, the damage can be in the form of crack or scratch due to the tool. The damage can give an impression that processing the top surface of a block by machining can have a negative result on the side surface as well; this can give erroneous conclusion that machining and AM are not basically different. The damage can be due to the lack of right tool or right tool speed or other conditions, but in an ideal case there will be no damage, and the surface quality will not depend on the tool. In AM, the side surface depends upon the top surface. Even in an ideal case, when the side surface will be the best, the formation of the side surface will not be free from its dependence on the top surface. This leads to conclude that machining and AM are basically different.

2.6   Machining Has Many Build Directions

In machining, any surface of a block can be worked upon while AM has mostly one build direction. Thus in machining, all surfaces, i.e., side surfaces, top surface, and bottom surface, shown by C, D, A, and B, respectively, in Fig. 5.5b, can be processed. While in AM, only one surface, i.e., top surface (A in Fig. 5.5a) can be processed. Other surfaces such as side surfaces (C, D in Fig. 5.5a) and bottom surface (B in Fig. 5.5a) cannot be processed.

If a design needs to be selected to be processed by either AM or machining, knowing the number of its possible orientations for processing helps find the right manufacturing type. In AM, the orientation of a design determines the speed, cost, quality, type of anisotropicity, and the fabricability of the design [10]. Orientation is important because during the fabrication, there mostly remains one orientation. There are many AM (such as PBF) where products are fabricated with only one orientation. If a design is complex, the design does not offer many orientations to choose from.

This is why machining has an advantage over AM. In machining, orientation of a block can be changed many times (depending upon the geometry) for processing. Any surface of a block can be processed depending upon the convenience. Moreover, orientation does not give rise to anisotropicity as it gives in AM [11]. Having many orientations for fabrication means more fabricability and an opportunity for better surface quality.

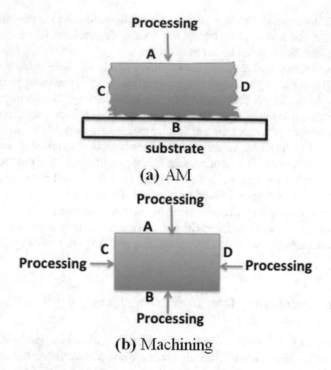

Fig. 5.5 Difference between possibilities of processing opportunities in AM (**a**) and machining (**b**)

3   Why Does Selection of Feedstock Influence Processing in AM More Than in Machining?

3.1   *Laser Powder Bed Fusion*

In LPBF, for example, if powder of the same material but of different shape is used, there is no guarantee that part properties furnished by both types of powder shape will be the same. For example, properties of parts made from two heaps of polyamide powder are different when each heap has different surface roughness.

For a heap of powders having high surface roughness, the attraction among powders will have minimum effect. If there is attraction among powders, they coalesce which decreases their flowability. Consequently, a heap of powders consisting of powders having high surface roughness will have better flowability than that having low surface roughness [12, 13].

Surface roughness thus affects how powders behave when they are laid on by a leveling roller or a scraper. If powders are of low surface roughness, they are not uniformly laid on giving rise to depression in the layer. When a beam scans the layer, these non-uniformities show up as non-uniform melt pools causing gaps between adjacent pools. This deteriorates the quality of the part. Thus if powders

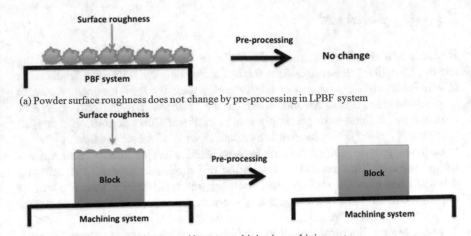

(a) Powder surface roughness does not change by pre-processing in LPBF system

(b) Block surface roughness is removed by pre-machining in machining system

Fig. 5.6 Pre-processing of feedstock in (**a**) LPBF and (**b**) machining

have low surface roughness, it adversely influences manufacturing. There is no technique that changes surface roughness during in-process stage (Fig. 5.6a), a beam cannot be applied to change the surface roughness by melting or eroding surfaces of individual powders. By changing the speed of the roller, the effect of the surface roughness cannot be mitigated. If powders are not carefully selected, there is no way the lack of selection in pre-process stage can be overcome during the in-process stage. Employing real-time control can mitigate the effect of the surface roughness, but this type of mitigation has limitation.

In machining, the difference in surface roughness does not have the same influence during the in-process stage (machining stage). For example, two polyamide blocks of different surface roughnesses are taken—if one block has higher surface roughness, this warrants an additional pre-step to fine machine the surface and change the surface roughness. The use of additional pre-machining helps both blocks to achieve the same surface roughness (Fig. 5.6b). Thus change in the surface roughness does not have any influence in the main manufacturing stage, i.e., machining, if an additional step is taken.

Consequently, if one property of feedstock is changed, it has an impact in AM while it has no impact in machining. Powders of the same composition have other properties besides surface roughness which cause difference between two heaps of powders. These other properties are powder size distribution, different powder shapes, humidity, and oxygen level of two heaps of powders, etc. Among many properties of powders, only surface roughness was chosen to compare feedstock in AM and machining. It is because for other properties of powders, there are no matching counterparts available in feedstock (polyamide block) for machining. Non-availability of matching counterparts in machining itself demonstrates the abundance of variability in the selection of same material during pre-process stage in AM.

3.2 Wire Based AM

If wire is of non-uniform diameter, the feed rate does not remain consistent giving rise to non-uniform deposition in the form of varied heights. If the internal diameter of wire feeder nozzle is large to accommodate the wire, the feed rate does not remain inconsistent. But, due to non-uniformity in the diameter, non-uniform amount of wire is fed. This non-uniform supply results in non-uniform deposition if applied beam energy is sufficient to melt the complete wire. If the applied energy is not high, the thinner section of the wire is overmelted while its thicker section is not melted resulting in a deposition consisting of unmelted material. The resulting deposited line consists of overmelted and unmelted components. Thus if a wire of non-uniform thickness is selected during pre-process stage, it has an adverse effect during the in-process stage.

In machining, if a rectangular block is selected which is not exactly rectangular, it has almost no effect during machining stage. The non-uniformity of the block requires an additional step of machining to transform it in a rectangular shape, afterwards, the next stage of machining is similar to that of the block. It demonstrates that the selection of materials during pre-process stage in wire based AM and machining has different consequences during the in-process stage.

It is a known fact in machining that the selection of different materials causes a difference in manufacturing, but it is not so known fact in AM that the selection of same material may cause differences in manufacturing.

4 What Is Small or Big Size? Which AM Process Is Better for Making a Big Metallic Part?

4.1 Small and Big

If there are three parts having size 2 cm, 15 cm, and 1 m as their biggest dimensions, it is easy to identify that they are small, medium, and big parts, respectively. It is again easy to identify the parts smaller than 2 cm as small and parts bigger than 1 m are big. But the problem starts when the part size is at the boundary of two types, for example, whether 20 cm size is medium or big.

To identify the size, resolution of the part can help. If the resolution is under 50 µm [14] (high resolution), the part is small. If the resolution is about 100–300 µm (low resolution), the part is of medium size. Thus, if there is a part of size 5 cm having resolution 50 µm, the part is small while the part of the same size but having resolution 200 µm is medium. A part having high resolution will contain smaller features than the part having lower resolution, therefore the same size part having higher resolution needs different post-process treatment than that having lower resolution. Similarly, if a part of size 20 cm is made with 100 µm resolution, the part is medium; if it is made with a resolution of 1 mm, the part is big because it has the characteristics of a big part. Such characteristics are absence of small features, conducive to post-processing for adding or removing some features, etc. Small parts are not conducive to such types of post-processing due to the limited option at such high resolution.

Thus, the resolution can help identify the size of a part if the size is at the boundary of the two types. This brings a question how a fixed boundary can be decided. If the question can be asked whether 20 cm size is big or medium, the question can also be asked whether 15 cm or 25 cm is big or medium. Whether a part is big or small depends more on the perspective and application unless the size is the extreme. Hence, unless there is a standardization on a fixed boundary, identification the size by the resolution is more to do with how a fabricated part should be treated or stored or post-processed than to suggest a process variant for the part of a given size.

Since the part is made by a process; there are processes, for examples, with high, medium, and low resolution. Hence, a process with high resolution makes small part. A process with resolution 50 μm makes parts of maximum size of some centimeter and is not expected to make a part of size 1 m. This brings a question what will happen if a process with high resolution will make a big part of some meters. It will not be practical because the completion of the part will take several months or years, and will not be cost-effective. There are other reasons as well. A process with high resolution is expected to make small features, and such a big part will not be having any of the features that will be small enough requiring such process. If the big part may have few small features, it will be more logical to search instead another method that can add small features on the big part. Moreover, such process with high resolution comes in a small system that has limited working area unable to accommodate a big part, if to be fabricated.

The difficulty in identifying the one size from another size is more due to the scalable nature of AM. A process working with 100–300 μm layer thickness can make parts starting from 2 cm continuing up to several meters. Thus, there is nothing in 2 cm part which will make it different from a 2 m part except its size and the effect of its size on properties. Hence, for parts made by the same process, the process does not give an identifying probe to distinguish between two sizes and hence there is no boundary from where one size type starts.

4.2   Resolution in AM

This brings a question what is a resolution. The resolution is the smallest increment possible in a part. For example, if the resolution is 50 μm, the smallest increment possible in the height of a wall is 50 μm. Figure 5.7 shows the smallest gap of 50 μm in the heights of two walls. The smallest increment possible in the height of the wall is the resolution. Since the small increment in the height is equal to the layer thickness in layer manufacturing, the resolution depends upon the layer thickness and will be either equal to or more than the layer thickness. Thus, if the layer thickness is low, the resolution will be high. Consequently, if minimum layer thickness

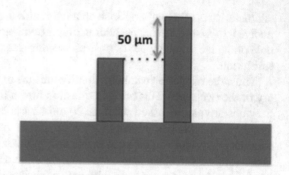

possible is 50 μm, the resolution cannot be 25 μm. Resolution will be 50 μm only when the layer after consolidation remains 50 μm. If after consolidation, the layer height changes such as somewhere it becomes 46 μm and some other place on the layer, it becomes 54 μm, the resolution will be 54 μm. Thus, if the consolidation does not happen accurately, the value of the resolution becomes higher (from 50 to 54 μm) and the resolution becomes lower.

Hence, it is the consolidation which decides what will be the resolution. The consolidation depends upon other factors depending upon the process. If it is a beam based process, the type and quality of beam, material type and their forms will affect the consolidation. If it is a non-beam based process, the type of binder, curing, material and their forms will affect the consolidation. But, the layer thickness is a dominating factor because it decides the boundary of the consolidation. If the layer thickness is 50 μm, the consolidation will not cause the thickness to change drastically at some place of the layer to 100 μm, for example.

The consolidation does not happen only in the vertical direction, it happens in the horizontal direction as well. In the horizontal direction, there is no limit set by the layer thickness, the consolidation go as far as possible if it is not controlled by controlling the parameters. Thus, there is not one resolution in AM. There are two resolutions: one mainly decided by the layer thickness, i.e., vertical resolution while other not mainly decided by the layer thickness, i.e., lateral resolution, also called spatial resolution [15]. If the process is well developed, the difference between two resolutions will be low. But if the process is not well developed, decreasing the layer thickness, even if machine allows, will not help get better resolution of the part.

4.3 Accuracy in AM

Thus the vertical resolution cannot be less than the layer thickness while the lateral resolution can be less than the layer thickness. Consequently, the size change of the part is limited in the vertical direction due to the layer thickness while is not limited by the layer thickness in the horizontal direction. It is limited by the accuracy of the consolidation in the horizontal direction.

If, in the horizontal direction, the increment is planned by 90 μm, and after measurement it is found that at some place along the consolidation line it is 94 μm while at other place, it is 86 μm, then the accuracy of consolidation is limited by 8 μm or ±4 μm. Thus, there is limitation in achieving consolidation or dimension by ±4 μm. This is the limitation of the process in the horizontal direction. This is the lateral resolution of the process. Thus the lateral accuracy and lateral resolution are the same thing.

What if an increment is planned by 90 μm in vertical direction. This is not possible since the increment must be multiple of the layer thickness. If the layer thickness is 50 μm, increments must be 50 μm, 100 μm, 150 μm, and so on. The luxury of having any increment in the horizontal direction cannot be imitated in the vertical direction. The increment in the horizontal direction is limited by the accuracy of the consolidation, if it is 8 μm, the increment cannot be less than 8 μm. But all increments above 8 μm can be achieved.

Thus, if an increment of 50 μm, i.e., equal to the layer thickness is planned in the vertical direction, this is the minimum increment in the vertical direction. If, due to the consolidation, there comes an error of 8 μm or ±4 μm in the height, the minimal increment is no longer than the actual minimal increment. The increment that was set as 50 μm is 50 + 4 μm = 54 μm after the consolidation. If the minimal increment cannot be equal to the layer thickness, the resolution cannot be equal to the layer thickness. The process has lost its higher resolution due to the inaccuracy in the consolidation, the resolution has become lower, the value of resolution is henceforth equal to the actual minimal increment, i.e., 54 μm. If machine allows to decrease the layer thickness, it does not imply that the higher resolution can be achieved without overcoming the problem of consolidation.

Thus in vertical direction, there is a resolution and there is an accuracy. The resolution depends upon the accuracy but the accuracy does not depend upon the resolution. Consequently, there is a vertical resolution and vertical accuracy while there is only one lateral resolution or accuracy. Thus a process can have the following determiner: vertical resolution, vertical accuracy, and lateral accuracy. The vertical accuracy and lateral accuracy can be same or different. It depends upon the process, there are instances when shrinkage in vertical and horizontal directions is different giving rise to different accuracies in both directions.

Thus, if a part having one feature of height 100 μm is planned to be made using a process having accuracy ±4 μm and resolution 54 μm, the size of the feature is expected to be within 96 μm and 104 μm due to inaccuracy (Fig. 5.8). If, on the part, it is planned to add another feature having minimum possible increase in size than the first feature, the size variation of the new feature will be within 146 μm and 154 μm due to the resolution. It is not possible to add another feature lower than 154 μm height.

Moreover, if material to be processed changes which brings a change in accuracy to ±6 μm, then the height of the feature of planned size 100 μm will be expected to be within 94 μm and 106 μm. If, on the part, it is planned to add another feature, the size variation of the new feature will be within 144 μm and 156 μm. This is because the resolution due to new accuracy will be 56 μm.

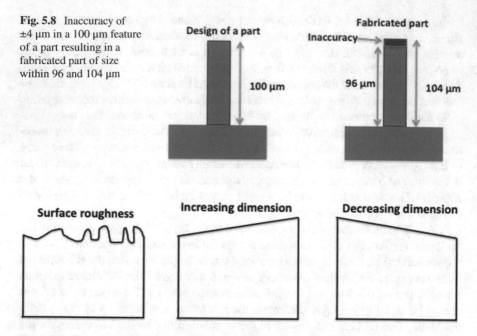

Fig. 5.8 Inaccuracy of ±4 μm in a 100 μm feature of a part resulting in a fabricated part of size within 96 and 104 μm

Fig. 5.9 Effect of uncertainty or inaccuracy on a fabricated part (not up to scale)

On the measurement of the height of the fabricated feature (Fig. 5.8), the height can be found to be 96 μm, 100 μm, 97 μm, etc., but, it will lie within 96 and 104 μm. Due to an uncertainty of ±4 μm, the upper topography of the feature changes from 96 μm to 104 μm. There will be bump, dent, and waviness on the upper surface but within ±4 μm.

If the process has no such uncertainty, the upper surface will not be such rough. This does not imply that uncertainty is only equal to the surface roughness. If this was the case, some surface treatment could be used, if possible, to remove the roughness and could nullify the effect of the uncertainty. Then, there will be no uncertainty any more. This type of uncertainty meant there would always be certainty but after doing some surface treatment.

The uncertainty means what is going to happen within ±4 μm is not certain. The height does not have certainty that it will always be 96 μm, 100 μm, 97 μm, etc.; the height may change in increasing pattern from 96 μm to 104 μm or in decreasing pattern from 104 μm to 96 μm (Fig. 5.9) or in some other patterns. Then, it will not be a problem related to surface roughness but a problem related to a change in dimension. There is thus no guarantee that a treatment for removing the surface roughness will bring back the inaccurate part to that condition that could be reached if the uncertainty meant only the problem of surface roughness.

4.4 Specification for Machine

Using this resolution and accuracy, a feature can be made having the height of 54 μm and a base having size 8 μm × 8 μm. This will be the minimum size that can be added on an existing part. But, whether a feature of this size can be made as a stand-alone feature or as a separate part. Making a separate part of this size and adding a feature of this size are two different things. In the former case, it depends on how it will be removed from a build plate, or how making a separate part will again affect its size and accuracy.

Thus, information about an AM machine is complete if it contains the following:

1. The smallest feature it can make as a separate feature.
2. The smallest feature a part can have as an add-on feature.
3. The vertical resolution of the product that will be formed. It is not about the positional accuracy of the machine or the vertical resolution of the machine. There is no guarantee that the resolution of the machine will be translated into the resolution of the product.
4. The vertical and horizontal accuracy of the product that will be formed.
5. The range of processing space of the machine within which the accuracy will work.

4.5 Accuracy and Repeatability

Figure 5.8 shows two parts fabricated are accurate within a range of ±4 μm. This is the accuracy of these parts. If many parts are fabricated, they all should come within the range of ±4 μm. If they come within the range, accuracy will not be different from repeatability; accuracy and repeatability will be the same.

But, determination of accuracy does not ensure that all parts will henceforth fall within the range. Determination of accuracy only means what the process has to offer in an ideal condition. Accuracy is the limitation of the process. Accuracy is silent how the process will be executed, it is also silent when and where the process will be executed. These questions will not be important if the system is robust enough to be immune from such sources of variability. But in practice, it does not happen so. If it happened, accuracy and repeatability would be the same.

In AM, there are a number of sources of variability. For example, what is the orientation of the design, what is the location of the design in the build plate, what is the location of the design in the processing space (how much above the build plate), which batch of material is getting used, which is the system, what is the model of the system, etc. (refer Chap. 3).

Thus the system can change accuracy. Considering a case where accuracy is not ±4 μm but +4 μm. Henceforth, the size of the feature will change positively. A feature designed to be 50 μm will become 54 μm, and a feature of 100 μm will become

104 µm and so on. Thus, due to the system, there is no more uncertainties of ±4 µm. Accuracy has become better. Henceforth, whatever size will be fabricated will become bigger by +4 µm. Thus, repeatability is better than accuracy.

Since it is certain that the size will change positively by +4 µm, this change will not be able to bring actual change in the size of the part. For making a feature of 100 µm, the feature will be designed to be 96 µm so that due to the positive accuracy of +4 µm, the designed feature of 96 µm will become 100 µm.

Thus a feature which needs to be designed 100 µm will be designed 96 µm after gaining knowledge how the system works. This is called offset in the design. The design is set 4 µm off from 100 µm. Thus the repeatability is determined by the system. In this case, the accuracy is superseded by the repeatability.

4.6   Why Big Part

AM is suitable for making small or medium sized complex parts and is not expected to make a non-complex large part. Large parts are usually made by CM such as forming, forging, casting, or machining. However, there are following reasons for which a large part can be made in AM:

1. When a part needs to be fabricated at the earliest while lead-time (time from design to fabrication) for the same in CM is high.
2. When there is a need to conserve materials such as tungsten [16], titanium, tantalum [17], niobium, etc., which can be wasted as chips if processed in CM [18].
3. When a new part is made, e.g., aircraft frames for aerospace or defense applications and secrecy needs to be maintained for either getting patent or preserving trade secret. AM requires few equipment to convert a design into a part, it is thus better than CM to maintain secrecy.
4. When the fabrication of a complex large part, e.g., a turbine blade or an engine component requires several CM sub-processes to complete while few processes (one main process plus one post-process) are required in AM.
5. When fabrication by CM is difficult [16].
6. When a part is less expensive to be fabricated in AM, e.g., when a single part needs to be made for which CM is more expensive due to the tooling or machining cost [18], or when the process uses low cost feedstock [19].

In principle, all metallic AM processes are capable to make large parts. In bed process (BP), a large bed is required, the size of the bed must not be less than the size of the part. To prevent oxidation, the bed needs to be enclosed in a chamber. The chamber also protects the outside environment from fine powders that become air-borne due to the processing.

4.7   Fabrication Time

For a big part to be completed in LPBF, it will take continuous operation of several weeks if scanned with high scan speed, several months if scanned with low scan speed. If a part of dimension 1 m × 1 m × 1 m is to be fabricated using 0.2 mm layer thickness, it requires 5000 layers to be scanned. If a layer is scanned using a hatch spacing of 0.2 mm and scan speed of 5 m/s, scanning beam has to traverse a distance of 5000 m which requires 1000 s (or 16.6 min). Thus for doing complete scanning of the whole part, beam has to scan 5000 layers that will take 5000 × 16.6 min = 83,333 min or approximately 57.8 days.

In this example, the time for spreading the layer is not taken into account. If spreading one layer takes even 1 min, the total time will be 5000 min. Thus the total fabrication time will be 57.8 days plus 5000 min. In this example, the part is assumed to be a solid block but the actual part may not be a solid block, it may be having some hollow sections that will not take any time to scan. Therefore, the scanning of an actual part having an outer dimension of 1 m × 1 m × 1 m will take less time than 57.8 days. Thus the total fabrication time including time for spreading layers will be less than 57.8 days plus 5000 min. Hence, for a part having several hollow sections, the fabrication time may not be much different from 57.8 days.

Scan speed is taken as 5 m/s. Higher than 5 m/s are available for experimentation, but the use of such higher speed is not a common practice. Generally, the speed is less than 1 m/s, but if this speed is taken, the total time for fabrication will be continuous operation of around 10 months. Since the part to be fabricated is big, lower scan speed 1 m/s will not be tried if 5 m/s works well.

Fabricating a big part in LPBF, the machine requires to work continuously which may generate enormous heat that needs to be controlled. Long continuous operation may deplete powder available within the machine, which will cause interruption and refilling. If the machine requires to be stopped, it will take more time than 57.8 days. The fabrication time of around 2 months shows that LPBF is a slow process and is not rapid manufacturing, its slowness becomes prominent when a part having big size is contemplated to be fabricated. This rate of fabrication is low and will not be able to fulfill all industrial requirements but will not be unable to fulfill some prominent applications (refer Chap. 8).

If the part is tried in EPBF, the fabrication time will not be much different. Commercial EPBF machine has scan speed around several thousand meter per second (8000 m/s), but this speed will not help increase the fabrication speed. This speed is a speed meant for a beam to jump from one point to another point on a bed and not to travel from one point to another point by making a line on the powder surface on the bed. This speed is thus utilized for creating simultaneous several melt pools at the bed to impart uniform heating and minimize thermal gradient. This speed is not what is used to make a solidified line by traversing along a line on the bed.

Assuming that this type of high speed is selected instead for scanning along a line on the bed, the powder along the line will not melt but ablate [20]. The scan speed has a limitation—there is a maximum speed that will work for fabrication, beyond that maximum speed even if higher speeds are available in a machine, these speeds will not help converting powder fast into a solid body. But this speed (8000 m/s) has importance—the importance is to jump from one point to another, if the jump is not fast, there will not be simultaneous creation of melt pools. This fast jump is not significant enough in the light of actual slow line scanning to increase the fabrication rate. Thus fabrication rates in LPBF and EPBF are not significantly different even due to high difference in scan speed mentioned in commercial LPBF and EPBF systems.

4.8 Limitation of Powder Bed Fusion

PBF has inherent limitation for getting scaled up. The layer thickness cannot be increased beyond a certain thickness. The problem of layer thickness is that it cannot be melted through the thickness completely beyond a certain thickness. The layer is supposed to be melted from irradiation at the upper surface; therefore, majority of heat to melt the layer comes from only one surface, i.e., upper surface. Since there is only one surface, the surface limits the distance up to which heat can go through and reach starting from the upper surface. Thus, the upper surface receives heat for longer duration than any horizontal section of the layer away from the upper surface. Hence, the upper surface is heated more. If the layer thickness is large, attempt to melt the lower surface of the layer will cause the upper surface to accumulate more heat than necessary for melting. This will lead upper surface to be evaporated. Thus, scaling up the applied heat to melt the large thickness is possible if the damage on the surface leading to a damage on the adjacent zones is acceptable. But this is not the case.

4.9 Solid Deposition Process

In DP, scaling up the process can happen by scaling up the amount of deposition that will lead to scaling up the layer thickness. If a thin metallic wire of 0.5 mm diameter is used to deposit a small amount of material to a small height to make a small part, a thick metallic wire of 1 cm diameter can be used to deposit a big amount of material giving rise to bigger layer thickness. If melting of still higher thickness of wire is possible, it is because melting of the wire is not restricted to be melted only through one side. The melting of a layer in PBF has no such freedom.

The freedom to melt any thickness of wire in DP is an advantage which can be used to make a layer of higher thickness. For making a thin layer, thin wire will be melted which will be deposited in several adjacent lines. Similarly, for making a thick layer, thick wire will be used. Thus the process can be scaled up. The disturbance that happens in PBF on going from thin layer to such thick layer does not happen in DP.

Though after a certain thickness of wire or after a certain amount of powder, it is not possible or convenient to melt in-situ, there may be problem in proper melting or there may be heat-induced difficulty in the system. But scaling up DP can still continue if there is an alternative arrangement for the higher amount of material to be melted and then deposited. For example, if there is a molten metal reservoir instead of a solid wire, the problem of melting the wire is not required. But the ability to control the deposition and microstructure that happens with wire or powder will not happen with molten material. Nonetheless, scaling up the process will not be prevented for the reason higher layer thickness is not practically impossible as in PBF.

Due to its limitation, PBF does not give high feedstock conversion rate; it is about 0.25 kg/h while the same for laser powder DP is about 0.5 kg/h. For wire DP, conversion rate or deposition rate is about from 2 kg/h to 10 kg/h, while the process that uses electron beam instead of laser beam reports higher deposition rate. Thus DP gives higher deposition rate than PBF and is better for making large metallic parts. In contrast, conversion in micro scale AM is about 63 mm^3/h for powders (about 0.6 g/h for iron powder having density 9.8 gcc) [21], 8000 mm^3/h for photopolymers while 0.02 mm^3/h in two-photon photopolymerization [14].

4.10 Wire Instead of Powder

Wire has certain advantage over powder. For example, DP using wire has higher deposition rate (so that bigger part will be made quickly) and material efficiency (so that most of the wire will become constituent of the part). Further, wire is less expensive than powder [22] and making a big part with wire instead of powder gives economic advantage and demonstrates as well that a small difference in feedstock price causes a large difference in the price for a big part. Besides, large part produced by wire consumes less power than that produced by powder based AM [23].

Wire instead of powder has a less chance to be spilled during feeding. A blown powder can be lost by deflecting from a substrate or a deposited structure. Lost powder is difficult to be recovered completely. Besides, it can cause pollution to the environment and becomes health hazard. Thus form a health point of view, a powder deposition system rather than a wire deposition system is required more to be enclosed in a chamber.

There are three main wire deposition processes based on three types of applied energies used: laser wire DP based on laser, electron beam wire DP based on electron beam and wire arc additive manufacturing (WAAM) based on arc [22]. A covering chamber is not required in laser wire DP and WAAM. To prevent oxidation during processing, shielding gas such as argon or helium can be applied on the site during melting of the wire. But in the case of electron beam wire DP, a vacuum chamber is necessary for an electron beam to function. Hence, a part needs to be fabricated within a chamber. For laser wire DP and WAAM, there is no such restriction of a chamber, there is more freedom to develop the process for fabricating a complex large part using either robot or gantry based system.

WAAM can make large parts having medium complexity [24] but if the part consists of small features, WAAM may not be suitable since laser powder DP has flexibility to use small diameter beam and powder to make small features while the arc does not give such precision.

5   How Composite Is Formed in AM?

5.1   What Is Composite

A composite material is made from two different materials such that the combination of two materials is able to achieve a property not achievable by either material. For example, if metal and ceramic is combined, metal provides ductility while ceramic provides strength. The resulting composite thus formed is both tough and strong. In the absence of metal, the ceramic alone will be brittle while in the absence of ceramic, the metal alone will be weak.

Composite works because in ceramic composite the load gets transferred from ceramic to the metal when the composite is subjected to force. This transfer of the load saves the composite from getting fractured. While in metal composite, the transfer of the load from metal to the ceramic saves the composite from being elongated. Thus the composite will work if there is enough transfer of the load due to enough presence of the material that is in minority (i.e., metal is minority in ceramic composite). Enough transfer of load means there requires uniform dispersion of the material. Enough presence of the material means the formation of matrix giving rise to metal matrix composite, ceramic matrix composite, etc. If the amount of material is low, enough transfer of load does not happen, the ceramic composite will fracture in the ceramic way while the metal composite will be elongated like a pure ductile metal.

If the amount of material is low, the resulting combination of two materials is still composite but it does not give the benefit of a composite. Consequently, such combination of two materials is a composite because it is not a compound or alloy or intermetallic or pure metal. But, such combination of materials is not a composite because it does not fulfill the criteria of the composite. In order to distinguish between two cases, one is called a metal–ceramic composite while the other is called metal matrix composite or ceramic matrix composite.

But, materials are not always added to facilitate the transfer of load. It is added to increase other properties such as thermal conductivity, electrical resistivity, heat resistance, etc. In that case, the resulting combination of materials is a composite not because it has achieved better mechanical properties but because it has achieved better non-mechanical properties.

5.2 *Extension of Concept*

Fabrication of metal matrix composite or ceramic matrix composite means successful transfer of load due to the presence of ceramic and metal within an area of few microns or few hundred microns. What if the load of transfer is desired not due to the presence of composite components within an area of few hundred microns but due to the presence of composite components within an area of few millimeters or few centimeters. If scale becomes bigger, the components of the composite also become bigger. In matrix composite, the particle is of the size nano meter or several microns. For increased scale, the components of the size of microns or hundred of microns will not work, the components of the size of millimeter or centimeter are required.

Since that big size of particles or fiber diameters is not available, hence they need to be made. When they are made, there comes a question what will be their properties. The composite is made to transfer the load, the transfer of load is relevant when one component needs to transfer the load while another component needs to receive the load. It implies that one component is weak while the other component is strong. Thus making bigger particle and fiber is not sufficient, they should have different strength so that they can be combined to make composite.

When the components are in bigger dimension and a common medium sized product is made, the size of the product is not much bigger than the components (Fig. 5.10a), then the shape of the component will start interfering in the accuracy of the shape of the product. Thus the bigger particle and fiber will not work for making the shape of all products. New shapes of the components need to be searched. When the components are in smaller dimension, their shapes are too small to affect the shape of the product (Fig. 5.10b). Therefore the selection of the shape of the component is done only on the basis of how it will affect the mechanical properties of the composite. But when the size of the component is bigger, this cannot be the only basis for the selection.

When the size of the component is bigger, it is expected to transfer the load at bigger scale and the resulting combination of bigger components will make a composite. This composite again will have the property of a composite, for example, lighter and stronger, not achievable by either component. But the component is big. When the component is big, it has the possibility to be made by a number of small components (Fig. 5.10c). If it is made up of smaller components, there is a possibility that these smaller components are again weak and strong, they are then combined to give a composite. This composite is due to the transfer of load in smaller scale.

Fig. 5.10 Schematic
diagrams of composite-1
and composite-2: (**a**)
composite structure made
from weak and strong
structures, (**b**) composite
made from weak and
strong materials, (**c**)
components of composite
structure are made from
composite materials

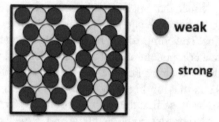

(a) Composite structure made from
weak and strong structures

(b) Composite made from weak and
strong materials

(c) Components of composite
structure is made from composite
materials

5.3   Composite-1 and Composite-2

Thus, there is a composite due to the transfer of load at bigger scale (composite-1)
and there is a composite due to the transfer of load at smaller scale (composite-2).
Composite-1 can be made by many materials, but it can be made by composite-2 as
well. Composite-2 can lead to many products but can also lead to various shapes
that will lead to composite-1 as well (Fig. 5.10).

Composite-1 is made by changing the structure at bigger scale and thus a com-
posite-1 part is made while composite-2 is either procured from somewhere to be
used in AM to make composite-2 part or composite-2 is developed in an AM system

while making composite-2 part. Examples of composite-1 parts are metallic sandwich structure [25] and continuous fiber reinforced sandwich structure [26]. Examples of composite-2 parts are WC-Co composite part by WC and Co [27], metal matrix composite using TiB_2 reinforced Al powder [28].

5.4 Composite-1

AM can make various shapes. These shapes can be made to be either strong or weak as well. The shape is not weak because AM cannot make strong shape. The shape is weak because strong shape is not required. AM gives opportunity to make weaker shapes by creating various degrees of hollowness. For example, if a shape is made up of lattice structure, the shape can have various strengths depending upon the size, type of the unit cell of the lattice, density and thickness of the strut of the unit cell, etc. Thus the weakness of a structure can be controlled and a weak structure of defined weakness can be made. For making a strong structure, a dense plate can be made without using the lattice. Making lattice or scaffold is not new in AM [29].

Consequently, weak structure is a lattice structure while strong structure is a dense plate (Fig. 5.11). If weak and strong structures are alternately changed, it makes a composite structure (Fig. 5.12). This structure is a sandwich structure where weak structure is called core while the strong component is called panel [25].

Strong-weak fabrication or panel-core fabrication has similarity with layer by layer fabrication. If one layer (or several layers) is core (weak) and another layer (or several layers) is panel (strong), pursuing layer by layer fabrication leads to an alternate arrangement of strong-weak (panel-core) that is called composite. Due to the similarity between panel by panel progression in composite and layer by layer progression during building, AM provides less complexity in process planning. Thus, developing such architecture of composite is not demanding from AM other than what AM is accustomed to offer.

If the composite shown in Fig. 5.12 is turned by 90° so that the direction of panel progression changes from z-direction to x-direction, the panel by panel development is not similar to layer by layer development. In that case, not a single layer or a consecutive group of layers will be dedicated to either core or panel only. For

Fig. 5.11 Schematic diagram of a sandwich structure

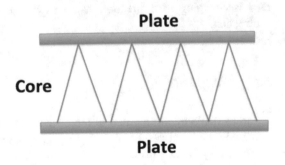

Fig. 5.12 Alternate fabrication of weak and strong structure in z-direction to make a composite

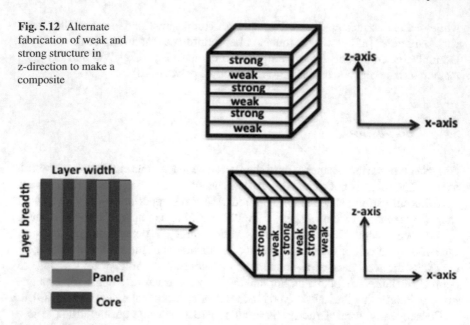

Fig. 5.13 Fabrication of a composite structure by arranging two different structures alternately in the same layer

Fig. 5.14 Methods to fabricate a composite structure part in AM

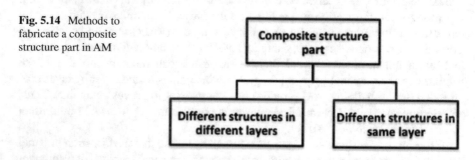

fabricating such composite, each layer will comprise of some part of each panel and core (Fig. 5.13). This shows there is another method for fabricating composite structure. These two methods are given in Fig. 5.14. In the first method, composite property is changing in z-direction while in the second method, it is changing in x-direction.

5.5 Composite-2

It can be made by mixing two different feedstock in a right combination so that a composite material results. Another method is to use different components of a feedstock to react during processing and create a composite material. Post-processing such as heat treatment, chemical reaction, infiltration is also used to

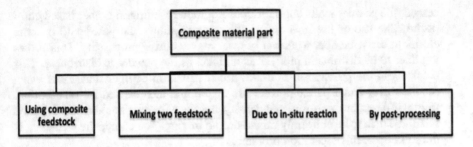

Fig. 5.15 Methods to fabricate a composite material part in AM

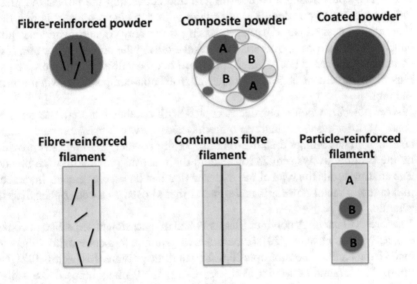

Fig. 5.16 Types of composite feedstock

make a composite material part (refer Chap. 4). Various methods to form a composite material part in AM are shown in Fig. 5.15.

The direct way to make a composite material part in AM is to use a composite feedstock. Other methods, without using composite feedstock, are to create composite material from non-composite feedstock.

5.6 Composite Feedstock

Composite feedstock means feedstock that consists of one material or one form type (Fig. 5.16). Examples are given below:

1. *Fiber-reinforced powder*: powder contains small fiber inside it. It is a composite because powder contains another material in the form of fiber. When it is pro-

cessed, the powder melts but fiber does not melt resulting in a fiber based composite. The use of the fiber inside the powder allows the powder to become carrier to carry the fiber wherever powder moves during processing. This allows the fiber to be distributed in layer as uniform as the powder is distributed. This helps avoid the problem of non-uniform distribution of fiber that arises when separate fiber and separate powder are mixed and then carried for distribution. Being different in size, form, and density, it is easy for the fiber and powder to be separated than to be uniformly mixed during mixing. The composite feedstock is the solution of this separation problem.

2. *Composite powder*: it is made from composite material. Methods to fabricate are mechanical alloying, melting, atomization, spraying, etc. One method to fabricate it is by pulverizing a composite part and converting the pulverized pieces into right sizes of powder. The advantage with composite powder is that their shaping in a desired part during processing is the only concern while it is not a concern whether a composite material results during the processing or not. If the composite powder is not selected, there are two concerns: shaping of the part, plus development of composite material from non-composite powders during shaping.

3. *Coated powder*: when a powder is coated with another material, the resulting coated powder has more than one material, necessary to form a composite. Since coating remains thinner than the diameter of the powder, the materials provided by the coating always remain lower than the material provided by the powder. The coating limits the type of the material that can be used as a coat. In the case of a ceramic-metal powder it is the metal that is used as a coat rather than the ceramic.

4. *Reinforced filament*: A polymer filament used in extrusion based AM can consist of small fibers, platelets [30], or particles to give rise to a composite. Fibers are made from carbon, thermotropic liquid crystalline polymer [31], glass [32], cellulose [33], aramid or Kevlar [34], etc. Using small fibers helps make a fiber-reinforced composite that is 30 to 40% stronger than a composite without reinforcement [35]. But the volume fraction of the small fiber in the filament always remains small which restricts the maximum strength that can be obtained. Replacing small fibers by continuous fibers helps increase the volume fraction of the fiber, which increases the strength [36].

   In layer by layer deposition, the job of the filament is finished when a layer is made. The job of the filament again starts when a new layer over the old layer is made. Thus the job of the filament is a discontinuous job. To maintain the discontinuity, when the deposition of a layer is finished, there should not remain any connection between the layer and the filament even through the extruded part of the filament. The extruded part of the filament becomes the part of the layer. Thus when the nozzle moves up, there is no more extruded part attached to the nozzle. Hence, the continuity of the extrusion gets broken when the layer deposition is over. If it does not break on its own, it needs to be broken in order to deposit the next layer.

When a filament reinforced with continuous fiber is used, the fiber does not get melted and is not supposed to be melted as well. Therefore, there is no chance that the continuity will be broken unless the fiber is broken. Hence, the fiber needs to be weak and the deposited layer needs to become quickly solid so that when the formation of the layer is over, upward movement of the nozzle is enough to break the fiber attached to the solid layer. This brings limitation to the diameter of the fiber (around 10 μm) and the number of fiber (around 1000) [37] that can be used in a filament [34]. In the absence of detachment of the strong fiber, post-processing is a solution to remove that section of the fiber or wire remaining between old and new layers, provided the layers can be deposited well without breaking the discontinuity.

It does not mean when a single layer needs to be made with continuous fiber, there will be no problem. Due to the stiffness of the continuous fiber, the circular movement of nozzle is restricted. When the nozzle needs to go back, it has to make a big circle rather than a small circle for returning back. Otherwise, the nozzle will not be able to make a sharp turn [31] (Fig. 5.17). This restricts the fineness of the feature that can be made.

5. *Composite pellet*: Filament is not suitable for all extrusion based processes, especially for making big parts. Thus, pellet is used. A composite pellet will help make not only a strong polymer composite part but also metal or ceramic composites which were not possible to be manufactured without using polymer to facilitate the processing. For example, a metallic alloy based composite magnet is formed using composite pellet [38]. This magnet got its shape because of the presence of polymer when the polymer facilitated metallic alloy to acquire the shape. The role of the polymer is more to give the magnet its shape than to change mechanical properties of the magnet by making a composite. Thus, polymer of the composite is a facilitator for processing than to give magnet a mechanical property that it cannot achieve on its own.

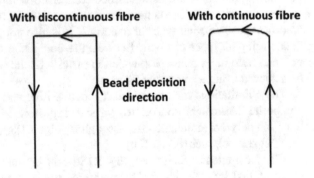

Fig. 5.17 Deposition of extruded material reinforced with discontinuous and continuous fiber, difficulty in turning due to continuous fiber

5.7   Bed Process

In powder bed process, if a metal composite powder is used instead of a metal powder, a part is formed which is made of only composite material. For example, if a composite powder made of Ti and TiC is used, a part is formed which consists of Ti-TiC composite, if a Ti Powder is used instead, a non-composite part, i.e., Ti part is formed [39, 40].

A single composite powder may consist of either two separate powders of Ti and TiC or a single Ti-TiC powder—in both cases, a composite part will result. If powder is fabricated due to different methods, the composite part will have different properties. In both cases the part formed will be called a composite part. Instead of taking a single composite powder (which is made of two separate powders), two separate powders can be taken without being part of a single composite powder— this can form a composite part but the presence of separate powder during spreading and consolidation thereafter may cause separation of two components of the composite, which will not help form a part having uniform distribution of composites; though, it gives freedom to make a composite from those materials which are not available in the form of composite feedstock.

For making iron-iron carbide composite, iron powder and graphite powder can be used. During processing, iron and graphite will react to form iron carbide. Thus a composite forms which consists of iron (that did not react with graphite) and iron carbide (due to reaction). If iron and graphite are enclosed in a single powder (Fig. 5.16), there is more uniform distribution of composite than when iron and graphite are not enclosed in a single powder (but in two different powders).

5.8   Fiber-Reinforced Composite in DP

For making continuous-fiber reinforced composite, composite filament having continuous fiber is used. The composite filament needs to be developed, which is not readily available for all types of materials. It is also not possible to make the composite using all types of fibers. Besides, the composite filament restricts the fiber-polymer ratio in a printed part, which in turn limits the range of properties that can be achieved using a composite filament.

These limitations can be avoided if separate fiber and polymer are used to make composite. This can be done in two ways: a single nozzle is used to deposit separate fiber and polymer simultaneously through it (Fig. 5.18a), separate nozzles are used for fiber and polymer (Fig. 5.18b).

When a single nozzle is used (Fig. 5.18a), the amount of deposited polymer can be controlled by changing the temperature, speed of nozzle, or nozzle diameter. When a big nozzle is used, the amount of polymer can be increased, this will increase polymer to fiber ratio. This will in turn decrease the amount of fiber in a

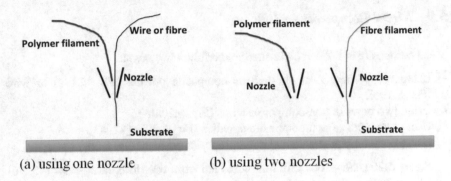

Fig. 5.18 Composite due to non-composite feedstock in solid DP: (**a**) using one nozzle, (**b**) using two nozzles

given volume of the part. Thus by changing the nozzle size, the volume fraction of fiber in a fiber-reinforced can be optimized. By changing the type of fiber and polymer, various composites can be formed for different applications. By replacing fiber with metallic wire, wire-reinforced composite can be formed for making mechanical or thermal sensor [41].

Single nozzle (Fig. 5.18a) in comparison to two nozzles (Fig. 5.18b) gives simplicity in its movement for deposition of two types of materials (polymer and fiber). But, single nozzle has limitation in increasing the volume fraction of fiber by changing the amount of polymer. Besides, the orientation of fiber in a layer is similar to the orientation of polymer bead. It cannot be different because the fiber intends to remain at the core of the bead.

Limitations posed by a single nozzle can be overcome by using two nozzles arrangement. With the help of two nozzles, the amount of fiber (volume fraction of fiber) added is no longer dependent on the amount of polymer. Their amounts can be independently varied. If their relative amounts are dependent, it is because they should provide right adhesion or property, it is not because extrusion of one needs to be related to the extrusion of the other.

Two nozzles allow polymer bead to be deposited at any orientation with respect to the deposited fiber. Thus the effect of various orientations on the property can be realized. Two nozzles allow one layer to be deprived of fiber while other layer to be deprived of polymer, thus the property can be further controlled. Depositing only fiber on a solidified layer will not work as a solid fiber will not attach to the solid layer. Thus a pure fiber must contain some glue to facilitate the bonding of the fiber to either an underlying solidified layer or an adjacent solidified bead. Commercial fibers though named as pure fibers contain some polymer such as epoxy to work in extrusion based system. Thus a fiber-reinforced composite made by two nozzles is not only the mixture of polymer and fiber but a mixture of polymer due to one nozzle and, glue and fiber due to the other nozzle.

5.9 Metal Composite in DP

Metal composite in DP can be formed in the following ways:

- Using composite feedstock such as composite powder (Fig. 5.19a) or wire (Fig. 5.19b),
- Using two types of non-composite wires (Fig. 5.20a).
- Using two types of non-composite powders (Fig. 5.20b) [42, 43].
- Using non-composite wire and non-composite powder (Fig. 5.20c) [44].

Figure 5.20a shows two different wires fed from two different nozzles [45, 46] which will melt in the presence of a heat source (a laser beam or an electron beam or a plasma beam or an arc, not shown in figure) to form a composite.

Different feed rates of wires will give different contributions of two wires in a resulting composite and thus different feed rates will lead to different properties of the resulting composites.

For example, two powder feeders one containing Ti powder and the other containing TiC powder can be set at different speed to make a number of parts each having different amount of Ti and TiC. Similarly, two wires made from Ti and TiC (one wire is Ti while other wire is Ti plus TiC) can be set at different feeding speed to make a number of parts each having different amount of Ti and TiC.

Thus a composite part is formed where each section of the part gets reinforced at the microstructural level. For example, TiC provides reinforcement and is

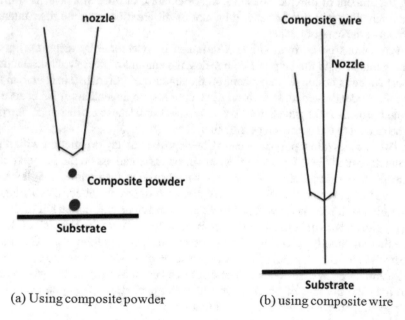

(a) Using composite powder   (b) using composite wire

Fig. 5.19 Fabrication of composite in solid DP using composite feedstock: (**a**) using powder, (**b**) using wire

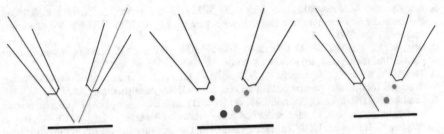

(a) using two different types of wire   (b) using two different types of powder   (c) using wire and powder of different types

Fig. 5.20 Fabrication of composite in solid DP using non-composite feedstock: (**a**) using wire, (**b**) using powder, (**c**) using wire and powder

responsible for increasing the strength of a pure Ti metal. Instead of TiC, other additives can be used for different purposes such as increasing electrical, thermal conductivity.

6 What Are the Extreme Variants of a Process?

AM processes are scalable [47], following are the examples of their extreme variants:

- In photopolymer bed process, there is mammoth stereolithography [48] for making meter-sized parts and microstereolithography [49] or nanostereolithography [50] for making millimeter-sized parts.
- In powder bed fusion, there is micro-selective laser melting [21, 51, 52] for making millimeter-sized parts and laser powder bed fusion for making parts of several centimeters [53].
- In extrusion based process, there is big area additive manufacturing [54] for making meter-sized parts and fused deposition modeling [55] for making millimeter-sized parts.
- In laser solid DP, there is metal big area additive manufacturing [24] for making meter-sized parts and micro-laser aided additive manufacturing [56] for making centimeter-sized parts.

References

1. Niaki, M., Nonino, F., Palombi, G., & Torabi, S. A. (2019). Economic sustainability of additive manufacturing: Contextual factors driving its performance in rapid prototyping. *Journal of Manufacturing Technology Management, 30*(2), 353–365.
2. Moshiri, M., Candeo, S., Carmignatio, S., et al. (2019). Benchmarking for laser powder bed fusion machines. *Journal of Manufacturing and Materials Processing, 3*(4), 85.

3. Antonysamy, A., Prangnell, P., & Meyer, J. (2012). Effect of wall thickness transitions on texture and grain structure in additive layer manufacturing (ALM) of Ti-6Al-4V. *Materials Science Forum, 205–210.*

4. Sow, M. C., Terris, T. D., Castelnau, O., et al. (2020). Influence of beam diameter on laser powder bed fusion (L-PBF) process. *Additive Manufacturing, 36,* 101532.

5. Tsai, C. Y., Cheng, C. W., Lee, A. C., & Tsai, M. C. (2019). Synchronized multi-spot scanning strategies for the laser powder bed fusion process. *Additive Manufacturing, 27,* 1–7.

6. Tenbrock, C., Kelliger, T., Praetzsch, N., et al. (2021). Effect of laser-plume interaction on part quality in multi-scanner laser powder bed fusion. *Additive Manufacturing, 38,* 101810.

7. Solyaev, Y., Rabinskiy, L., & Tokmakov, D. (2019). Overmelting and closing of thin horizontal channels in AlSi10Mg samples obtained by selective laser melting. *Additive Manufacturing, 30,* 100847.

8. Kowsari, K., Zhang, B., Panjwani, S., et al. (2018). Photopolymer formulation to minimize feature size, surface roughness, and stair-stepping in digital light processing-based three-dimensional printing. *Additive Manufacturing, 24,* 627–638.

9. Xiao, X., & Joshi, S. (2020). Process planning for five-axis support free additive manufacturing. *Additive Manufacturing, 36,* 101569.

10. Zhang, Y., Bernard, A., Harik, R., et al. (2017). Build orientation optimization for multi-part production in additive manufacturing. *Journal of Intelligent Manufacturing, 28,* 1393–1407.

11. Kok, Y., Tan, X. P., Wang, P., et al. (2018). Anisotropy and heterogeneity of microstructure and mechanical properties in metal additive manufacturing: A critical review. *Materials and Design, 139,* 565–586.

12. Ruggi, D., Lupo, M., Sofia, D., et al. (2020). Flow properties of polymeric powders for selective laser sintering. *Powder Technology, 370,* 288–297.

13. Schmidt, J., Dechet, MA., Bonilla, JSG. et al. (2019). Characterization of polymer powders for selective laser sintering. In *SFF Proceedings* (pp. 779–789).

14. Behera, D., Chizari, S., Shaw, L. A., et al. (2021). Current challenges and potential directions towards precision microscale additive manufacturing—Part II: Laser-based curing, heating, and trapping processes. *Precision Engineering, 68,* 301–318.

15. Singh, M., Kasper, F. K., & Mikos, A. G. (2013). Tissue engineering scaffolds. In B. D. Ratner, A. S. Hoffman, F. J. Schoen, & J. E. Lemons (Eds.), *Biomater Science* (pp. 1138–1159). Amsterdam: Academic Press.

16. Marinelli, G., Martina, F., Ganguly, S., & Williams, S. (2019). Development of wire + arc additive manufacture for the production of large-scale unalloyed tungsten components. *International Journal of Refractory Metals and Hard Materials, 82,* 329–335.

17. Marinelli, G., Martina, F., Ganguly, S., & Williams, S. (2019). Microstructure, hardness and mechanical properties of two different unalloyed tantalum wires deposited via wire + arc additive manufacture. *International Journal of Refractory Metals and Hard Materials, 83,* 104974.

18. Dominguez, L. A., Xu, F., Shokrani, A., et al. (2020). Guidelines when considering pre & post processing of large metal additive manufactured parts. *Procedia Manufacturing, 51,* 684–691.

19. Hassen, A. A., Noakes, M., Nandwana, P., et al. (2020). Scaling up metal additive manufacturing process to fabricate molds for composite manufacturing. *Additive Manufacturing, 32,* 101093.

20. Schleifenbaum, H., Meiners, W., Wissenbach, K., & Hinke, C. (2010). Individualized production by means of high power selective laser melting. *CIRP Journal of Manufacturing Science and Technology, 2*(3), 161–169.

21. Roy, N. K., Behera, D., Dibua, O. G., et al. (2019). A novel microscale selective laser sintering (μ-SLS) process for the fabrication of microelectronic parts. *Microsystems & Nanoengineering, 5,* 64.

22. Cunningham, C. R., Flynn, J. M., Shokrani, A., et al. (2018). Invited review article: Strategies and processes for high quality wire arc additive manufacturing. *Additive Manufacturing, 22,* 672–686.

23. Bekker, ACM., Verlinden, JC., & Galimberti, G. (2016). Challenges in assessing the sustainability of wire + arc additive manufacturing for large structures. In: *SFF Proceedings* (pp. 406–416).

24. Greer, C., Nycz, A., Noakes, M., et al. (2019). Introduction to the design rules for metal big area additive manufacturing. *Additive Manufacturing, 27*, 159–166.

25. Bühring, J., Nuño, M., & Schröder, K. U. (2021). Additive manufactured sandwich structures: Mechanical characterization and usage potential in small aircraft. *Aerospace Science and Technology, 111*, 106548.

26. Hou, Z., Tian, X., Zhang, J., & Li, D. (2018). 3D printed continuous fibre reinforced composite corrugated structure. *Composite Structures, 184*, 1005–1010.

27. Kumar, S. (2018). Process chain development for additive manufacturing of cemented carbides. *Journal of Manufacturing Processes, 34*, 121–130.

28. Xie, X., Ma, Y., Chen, C., et al. (2020). Cold spray additive manufacturing of metal matrix composites (MMCs) using a novel nano-TiB2-reinforced 7075Al powder. *Journal of Alloy Compound, 819*, 152962.

29. Tao, W., & Leu, MC. (2016) Design of lattice structure for additive manufacturing. In *International Symposium on Flexible Automation (ISFA)* (pp. 325–332).

30. Caminero, M. Á., Chacón, J. M., García-Plaza, E., et al. (2019). Additive manufacturing of PLA-based composites using fused filament fabrication: Effect of graphene Nanoplatelet reinforcement on mechanical properties, dimensional accuracy and texture. *Polymers, 11*(5), 799.

31. Ansari, M. Q., Redmann, A., Osswald, T. A., et al. (2019). Application of thermotropic liquid crystalline polymer reinforced acrylonitrile butadiene styrene in fused filament fabrication. *Additive Manufacturing, 29*, 100813.

32. Carneiro, O. S., Silva, A. F., & Gomes, R. (2015). Fused deposition modeling with polypropylene. *Materials and Design, 83*, 768–776.

33. Wang, Z., Xu, J., Lu, Y., et al. (2017). Preparation of 3D printable micro/nanocellulose-polylactic acid (MNC/PLA) composite wire rods with high MNC constitution. *Industrial Crops and Products, 109*, 889–896.

34. Dickson, A. N., Barry, J. N., McDonnell, K. A., & Dowling, D. P. (2017). Fabrication of continuous carbon, glass and Kevlar fibre reinforced polymer composites using additive manufacturing. *Additive Manufacturing, 16*, 146–152.

35. Parandoush, P., & Lin, D. (2017). A review on additive manufacturing of polymer-fiber composites. *Composite Structures, 182*, 36–53.

36. Werken, N. V. D., Koirala, P., Ghorbani, J., et al. (2021). Investigating the hot isostatic pressing of an additively manufactured continuous carbon fiber reinforced PEEK composite. *Additive Manufacturing, 37*, 101634.

37. Tian, X., Liu, T., Yang, C., et al. (2016). Interface and performance of 3D printed continuous carbon fiber reinforced PLA composites. *Composites. Part A, Applied Science and Manufacturing, 88*, 198–205.

38. Li, L., Tirado, A., Nlebedim, I., et al. (2016). Big area additive manufacturing of high performance bonded NdFeB magnets. *Scientific Reports, 6*, 36212.

39. Gu, D., Hagedorn, Y. C., Meiners, W., et al. (2011). Nanocrystalline TiC reinforced Ti matrix bulk-form nanocomposites by selective laser melting (SLM): Densification, growth mechanism and wear behavior. *Composites Science and Technology, 71*(13), 1612–1620.

40. Zhang, L. C., & Attar, H. (2016). Selective laser melting of titanium alloys and titanium matrix composites for biomedical applications: A review. *Advanced Engineering Materials, 18*, 463–475.

41. Ibrahim, Y., Melenka, G. W., & Kempers, R. (2018). Additive manufacturing of continuous wire polymer composites. *Manufacturing Letters, 16*, 49–51.

42. Syed, W. U. H., Pinkerton, A. J., Liu, Z., & Li, L. (2007). Single-step laser deposition of functionally graded coating by dual 'wire–powder' or 'powder–powder' feeding—A comparative study. *Applied Surface Science, 253*(19), 7926–7931.

43. Bandyopadhyay, A., & Heer, B. (2018). Additive manufacturing of multi-material structures. *Materials Science & Engineering R: Reports, 129*, 1–16.
44. Teli, M., Klocke, F., Arntz, K., et al. (2018). Study for combined wire + powder laser metal deposition of H11 and niobium. *Procedia Manufacturing, 25*, 426–434.
45. Wang, J., Pan, Z., Wang, Y., et al. (2020). Evolution of crystallographic orientation, precipitation, phase transformation and mechanical properties realized by enhancing deposition current for dual-wire arc additive manufactured Ni-rich NiTi alloy. *Additive Manufacturing, 34*, 101240.
46. Cai, X., Dong, B., Yin, X., et al. (2020). Wire arc additive manufacturing of titanium aluminide alloys using two-wire TOP-TIG welding: Processing, microstructures, and mechanical properties. *Additive Manufacturing, 35*, 101344.
47. Vaezi, M., Seitz, H., & Yang, S. (2013). A review on 3D micro-additive manufacturing technologies. *International Journal of Advanced Manufacturing Technology, 67*, 1721–1754.
48. Salonitis, K. (2014). Stereolithography. In S. Hashmi, G. F. Batalha, C. J. Van Tyne, & B. Yilbas (Eds.), *Comprehensive materials processing* (pp. 19–67). Amsterdam: Elsevier.
49. Han, D., Yang, C., Fang, N. X., & Lee, H. (2019). Rapid multi-material 3D printing with projection micro-stereolithography using dynamic fluidic control. *Additive Manufacturing, 27*, 606–615.
50. Suzuki, Y., Tahara, H., Michihata, M., et al. (2016). Evanescent light exposing system under nitrogen purge for nano-stereolithography. *Procedia CIRP, 42*, 77–80.
51. Abele, E., & Kniepkamp, M. (2015). Analysis and optimization of vertical surface roughness in micro selective laser melting. *Surface Topography: Metrology and Properties, 3*, 034007.
52. Nagarajan, B., Hu, Z., Song, X., et al. (2019). Development of micro selective laser melting: The state of the art and future perspectives. *Engineering, 5*(4), 702–720.
53. Bosio, F., Shen, H., Liu, Y., et al. (2021). Production strategy for manufacturing large-scale AlSi10Mg components by laser powder bed fusion. *JOM, 73*, 770–780.
54. Roschli, A., Gaul, K. T., Alex, M., et al. (2019). Designing for big area additive manufacturing. *Additive Manufacturing, 25*, 275–285.
55. Monzón, M. D., Gibson, I., Benítez, A. N., et al. (2013). Process and material behavior modeling for a new design of micro-additive fused deposition. *International Journal of Advanced Manufacturing Technology, 67*, 2717–2726.
56. Bi, G., Sun, C. N., Chen, H. C., et al. (2014). Microstructure and tensile properties of superalloy IN100 fabricated by micro-laser aided additive manufacturing. *Materials and Design, 60*, 401–408.

Chapter 6
Application

1 What Are the Applications of AM at Various Places?

There are many applications of AM at a place. Following are some of the applications:

- In office as a desktop to make a prototype [1],
- In a small workshop to make replacement parts [2],
- In a factory to make prototype and speed up the production process [3],
- In a service center for production [4],
- In a hospital to make a surgery tool [5],
- In a primary school for making toys [6],
- In a high school for making educational tools [7, 8],
- In a college for making engineering design [9] and part [10],
- In a studio to make an artistic part,
- At a heritage site for making souvenir [11],
- In a bus as a school [12],
- In a design studio to make a verification part,
- In an airplane for making an emergency part,
- In an architecture department to make a small replica of a building [13],
- In a jewelry shop to make a fashionable ornament [14],
- In a bakery shop to make a cake of any design using new food mixture [15],
- In Moon for using its dust to make parts [16],
- In a space station for making spare parts [17],
- In a service supply chain for making inventories [18].

© The Author(s), under exclusive license to Springer Nature Switzerland AG 2022
S. Kumar, *Additive Manufacturing Solutions*,
https://doi.org/10.1007/978-3-030-80783-2_6

2   What Is Repair and Refurbishment in AM?

Repair means bringing a damaged part to its original condition, it is mostly used for improving the surface but it may require some features to be built as well. Refurbishment means adding some features on the damaged part so that the part can become better than its original condition in some or all respects. Therefore, refurbishment is an opportunity to add new functionalities in an old part.

Figure 6.1 shows schematic diagrams of a part before and after damage. One of the features of the original part (Fig. 6.1a) is lost in the damaged part (Fig. 6.1b). AM is used to build the lost feature. Figure 6.2a shows the repaired part in which repaired feature is shown by red color. Thus repaired part is same as the original part (Fig. 6.1a). In repairing, AM did not change the design of the original part but remanufactured the damaged part in such a way that the damaged part becomes similar to the original part [19]. Repair also includes those parts (Fig. 6.2b) which cannot be brought back to the original condition and hence the effect of damage is not completely gone. AM cannot repair all damaged parts.

Figure 6.3a shows the refurbishment of the same damaged part. AM added extra feature on the damaged part so that refurbished part is no longer similar to the original part (Fig. 6.1a) but has become better. Thus refurbishment gives opportunity to check whether the design of an original part can be improved during repair. AM with improved capability in manufacturing can change the original part which was

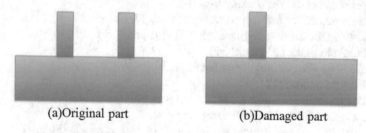

(a)Original part   (b)Damaged part

Fig. 6.1   Damage of a part (**a**) original part, (**b**) damaged part

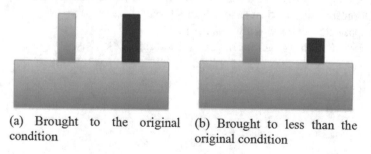

(a) Brought to the original condition   (b) Brought to less than the original condition

Fig. 6.2   Repaired parts (**a**) brought to the original condition, (**b**) brought to less than the original condition

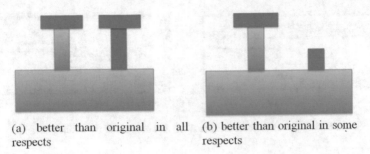

(a) better than original in all respects

(b) better than original in some respects

Fig. 6.3 Refurbished parts: (**a**) better than original in all respects, (**b**) better than original in some respects

not better because when it was made, the application of AM might not be known, or AM was not developed.

In case a damaged part cannot be repaired and brought back to the original condition, some new features in the damaged part can be added to compensate the lack of repair. Figure 6.3b shows a refurbished part which cannot be brought back to the original condition, but some new functionality is given to the damaged product by adding some features. The part does not become original-like but better than the original in some respects.

Thus remanufacturing by AM in the form of either repair or refurbishment not only salvages the damaged part but also improves the functioning of the part. While manufacturing in AM is done solely by AM, the remanufacturing is seldom carried out solely by AM. Damaged surface first needs to be prepared by machining to check the fittability of the part for remanufacturing and to maintain the accuracy of addition by AM [20].

If accuracy after AM is not right, it can again be machined. Machining can be planned to occur several times—this will lead to mix additive repair with subtractive repair [21]. This is again contribution of AM in remanufacturing.

Besides having design advantage, AM has material advantage over CM for repair and refurbishment. When, for example, a turbine blade needs to be repaired and the repair must provide only a single crystal (instead of multi-crystal) growth on the damaged blade, CM such as welding finds it difficult to prevent the growth of more than one crystal while PBF provides a single crystal growth [22]. Thus remanufacturing done in PBF provides both design and material solutions together [23], which traditional welding cannot match.

3 What Is the Advantage of AM for Making a Conformal Cooling Channel?

A conformal cooling channel is a cooling channel which is equidistant from the contour of a cavity so that the cooling of the cavity will not be affected due to the lack of proximity of the channel to the cavity.

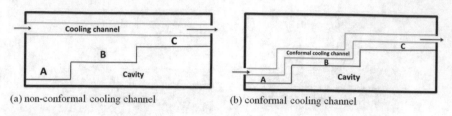

(a) non-conformal cooling channel   (b) conformal cooling channel

Fig. 6.4 Schematic diagrams of non-conformal (**a**) and conformal (**b**) cooling channel

Figure 6.4a shows a cooling channel which is at different distance from plane A, B, and C of the cavity. The cooling due to channel on the various areas of the cavity is not same. Figure 6.4b shows a cooling channel which is not a straight channel like shown in Fig. 6.4a but a channel which is bent twice to follow the contour of the cavity so that the distance of the channel from all three planes A, B, and C are same. This channel is a conformal cooling channel while the straight channel shown in Fig. 6.4a is a non-conformal cooling channel [24].

3.1 Channel by Drilling

Drilling makes a straight hole or channel, if a curved hole or channel needs to be fabricated, it is not possible. However, if no tool is available other than a drilling tool, an approximate non-straight or curved channel can be fabricated by drilling a block from various locations and orientations, and filling up extra gaps [25].

If a cooling channel (white in color, Fig. 6.5a) needs to be made in a metallic block (black in color, Fig. 6.5a), it requires two steps. The first step is to drill the block, which is shown in Fig. 6.5b by four arrow marks at four locations of the block. The second step is to fill up the extra drilled channel by filler, as shown in Fig. 6.5c by red color.

If the channel is a conformal channel which is curved in shape and is not of straight line type, drilling the hole and filling up the extra hole are not able to make the channel unless a number of tools is used to make holes a number of times.

3.2 Difficulty

If the cooling channel has variable diameters, one drilling tool is not sufficient. Many tools of various diameters are required. Arranging those tools, planning with them, setting fixture and orientation associated with the tools will be required. If a block to be drilled upon has some other features, drilling from all orientations without damaging the features may not be possible. If the diameter of the channel is frequently increasing and decreasing, even using tools of various diameters may

Fig. 6.5 Fabricating a cooling channel in a metallic block using drilling: (**a**) design of a cooling channel in a metallic block, (**b**) drilling at four places from two sides of the block, (**c**) filling up the extra drilled hole by fillers at four openings

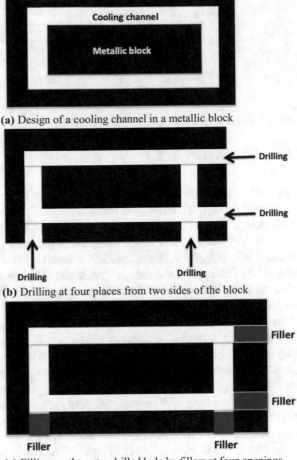

(**a**) Design of a cooling channel in a metallic block

(**b**) Drilling at four places from two sides of the block

(**c**) Filling up the extra drilled hole by fillers at four openings

pose difficulties. If the design of the channel is such that the opening of the channel has a small diameter while the diameter inside the bulk is large, drilling is possible only by breaking the block into two. The broken block needs to be joined afterwards. The breaking is not preferable but there is no other way to make such type of channels using drilling.

Though drilling cannot make a curved channel but vertical channel along with the horizontal cooling channel can help reach the contour of a cavity [26]. Figure 6.6 shows a channel consisting of vertical plus horizontal channels, which is better than non-conformal cooling channel (Fig. 6.4a) but is not better than conformal cooling channel (Fig. 6.4b). In the absence of free flow (Fig. 6.6), there will be limited benefit from these channels.

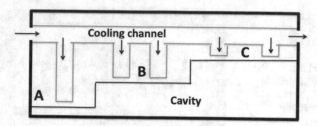

Fig. 6.6 Fabricating a cooling channel using drilling

3.3   *Channel by AM*

Difficulties by drilling show if a complex product having conformal channel can be fabricated by AM, it cannot be underestimated.

In AM, a channel is made by leaving the area within the channel unprocessed. In laser powder bed fusion (LPBF), a channel is made by melting all the area around a channel but not melting the area inside the channel. Melting around the channel gives rise to the diameter of the channel while unmelted or unprocessed area inside the channel gives hollowness of the channel. A bigger or smaller channel diameter can be made by leaving, respectively, the bigger or smaller area unprocessed. Hence, there is no more need to work with different diameters of drilling tools (refer Chap. 5).

Fabricating a narrow channel is possible if it is possible to leave narrow area unprocessed. For this, a beam of defined spot size is required to make small melt pool. The effect of the spot size on a given layer thickness at a certain scan speed is required to make an unmelted line. If the unmelted line or unmelted channel is narrow, the surrounding area should have low surface roughness. Otherwise, the narrow channel will become more narrow—this may not facilitate unmelted powder to be drained out. If this type of channel is compared with that made by drilling, it is clear that the channel made by drilling tool has better surface finish, accuracy, and high definition.

This shows that AM gives advantages to create a narrow channel at difficult locations of a design which is not possible by drilling. Though to get high definition of the channel as achieved with drilling tool, AM needs to go through optimization. Thus AM makes a product consisting of a complex channel. The position, size, and shape of the channel do not affect the fabrication of the channel in AM in the same way as they affect in drilling. But if a complex channel is not required, drilling is better because it provides better quality.

3.4   *By Investment Casting*

Drilling is not the only CM that is used to make channels. Casting [27] or investment casting (IC) is another CM used to make products having complex internal channels. This process is traditionally used to make channels before the advent of AM. The notable example is a turbine blade with internal cooling channel.

IC requires a number of steps to make a part. First, a duplicate part made from low melting point material such as wax is made—this is called pattern. The aim is to convert this duplicate part (pattern) into a real part using required material. To convert the pattern into a part, a negative impression from the pattern is required. The negative impression is filled up with the required material, this filled up material becomes a real part.

For making negative impression from the pattern, liquid ceramic is poured surrounding the pattern. When liquid ceramic is solidified and ceramic shell is formed partially covering the pattern, ceramic shell plus pattern is then heated so that the pattern being of low melting point material drains out of the shell. This shell is called mold. The hollow shell thus formed is ready to be filled up with metallic material. When the material is dried, the shell is broken to let the dried material out. This dried material is the desired part.

Thus for making a complex part from IC, a complex pattern needs to be fabricated. If the part is more complex and the fabrication of a complex pattern is not possible, many simpler patterns need to be fabricated and arranged to give rise to a complex pattern. Such type of arrangement will compensate the need for making a complex pattern. The arrangement should be enough robust that during pouring of ceramic liquid, the arrangement do not shift.

3.5   Advantage of AM

Pattern is used one-off—it implies that after making one part the pattern cannot be reused. If a number of parts of the same design needs to be made, a permanent pattern is fabricated which is used to replicate several wax patterns. Thus the process takes a number of steps to make a complex part while AM is a direct process and is therefore a faster process.

IC involves a number of steps, each step is prone to cause inaccuracy. Fabrication of a pattern by CM has limitations. These limitations are overcome by using AM. The limitations are—pattern wax needs to be drained out from the shell, the process may leave residue which may react with casted liquid metals to form defects. Moreover, the surface finish of a metallic part depends upon the formation of smooth ceramic shell. Setting time and selection of binder for ceramic slurry are required to fabricate shell having optimized properties. Casting and thereafter its removal by breaking the shell limit the fine features that can be fabricated.

Thus if a complex channel is not formed by AM, it is because AM may not be well developed for a particular material, not because AM has to pass through many dependable steps as IC has. Processing in IC depends on materials (pattern, shell) which do not end up in a product but get wasted, which makes IC expensive. AM processed part has better mechanical properties than casted part [28]. Thus AM has an edge over IC not only in the complexity but also in the mechanical property of a product that can be achieved.

4 What Are the Applications of AM in Casting?

AM is used to make permanent mold, temporary mold, permanent pattern for making mold, permanent pattern for making temporary pattern, and temporary pattern.

4.1 Permanent Mold

AM is used to make metallic or ceramic mold that can be used to make cast parts several times (Fig. 6.7). AM allows to make lattice structure in the wall of the mold for stress management and cooling channels for thermal management during die casting [29].

4.2 Temporary Mold

A complex part is difficult to be removed from mold. AM is used to make temporary mold such as sand mold [30] so that complex parts can be removed by breaking the mold which is not further reused (Fig. 6.8).

Sand mold fabricated by AM provides higher surface finish, better mechanical properties, and higher saving of sands than that fabricated by CM [31].

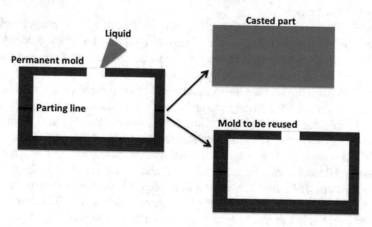

Fig. 6.7 Permanent mold that can be fabricated by AM

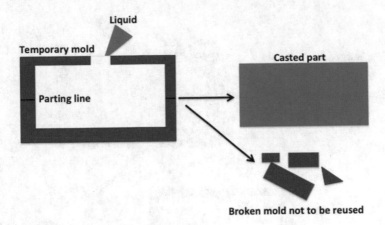

Fig. 6.8 Temporary mold that can be fabricated by AM

4.3   Permanent Pattern for Making Mold

AM is used to make plastic patterns that are further used for making permanent or temporary molds. These molds are then used for casting. Figure 6.9 shows a permanent pattern that is inserted inside a mold box where metallic liquid or ceramic slurry is poured over it. The liquid after drying takes the external shape of the pattern—this is the mold (from the pattern) which is used for casting. Thus the shape and surface roughness of the internal cavity are the result of the shape and surface roughness of the pattern while the external shape and the surface quality of the mold are decided by the inner volume of the mold box. The external shape thus formed does not directly affect the quality of part that is made by casting from the mold.

4.4   Permanent Pattern for Making Temporary Pattern

If a number of parts is required to be fabricated by, for example, investing casting, then instead of making a number of temporary patterns or wax patterns, a permanent pattern can be fabricated that can be used to replicate temporary patterns. AM is used to make plastic patterns which are further used for making permanent or temporary molds. These molds are used to make wax patterns. The wax pattern is used to make sand mold which is used for one-off casting. Making a temporary pattern from a permanent pattern through a temporary mold is shown in Fig. 6.10.

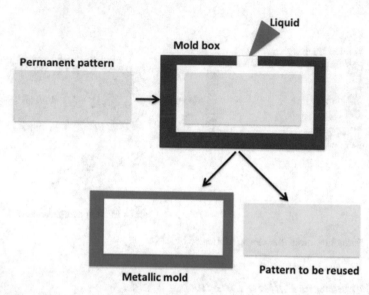

Fig. 6.9 Permanent pattern that can be fabricated by AM

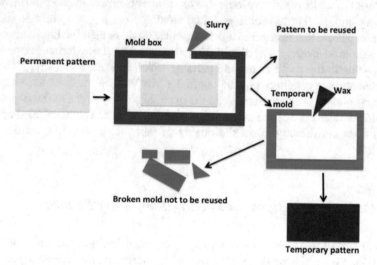

Fig. 6.10 Permanent pattern fabricated by AM, used for making temporary pattern

4.5 Temporary Pattern

If only one part needs to be made from casting, it is better to make a temporary pattern than a permanent pattern. AM is used to make temporary or expendable patterns using materials such as acrylonitrile butadiene styrene, polymethyl methacrylate, wax, epoxy, etc.

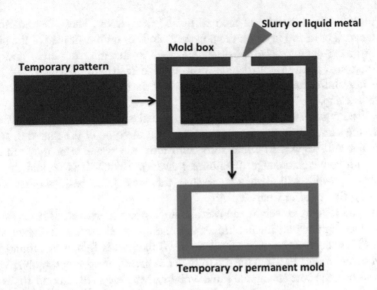

Fig. 6.11 Temporary pattern fabricated by AM, used for making mold

The pattern expands during removal of the pattern by burning, the expansion can break the solid cast material on it. To control the thermal expansion, the pattern can be made of hollow structure or internal lattice structure. AM allows to make such types of lattice structures and to control the stress generated by expanded patterns. As the patterns are expandable, the lattice also allows to minimize material waste by using less material for pattern fabrication [32].

The patterns are further used for making temporary or permanent molds. Figure 6.11 shows a route to make a mold using temporary patterns. For example, AM can make a lattice pattern, which can be infiltrated with plaster to make mold. Thus casting by aluminum on that mold will provide a regular lattice structure [33].

5 How Is Energy Saving Related to Material and Sustainability?

5.1 Energy Saving and Material

Producing 1 kg of iron powder takes approximately from 7 MJ to 15 MJ of energy, depending upon how they are produced [34]. In case powder is spherical and having narrow size distribution, producing 1 kg will consume more energy not less than approx. 10 MJ.

If AM is used to make a part weighing 1 kg, fabrication will require around 40 kg powder. Actual amount depends upon the type of a system and the density of the material. Remaining 39 kg powder will develop some form of degradation due to being present in the vicinity of the part during fabrication. These degradations are shape change, oxidized layer on the powder, satellites on the surface of the powder, changes in the chemical composition, changes in the microstructure and phase. Shape and size change will change powder size distribution which will in turn change flowability and spreadability of powder during building up powder layers [35] (refer Chap. 5).

The used powder having different characteristics than the unused powder will not give the same part property. Depending upon the type of the material, some of the powder will be wasted while some will turn into waste after they are reused (along with some percentage of unused powder). Hence, it is certain that some amounts of powder will always be wasted, this wastage is besides other wastage such as a spillage (refer Chap. 2).

If the part is light in weight and used in automobile or aircraft, it helps save fuel energy since lighter part means lighter automobile or aircraft as a lighter vehicle consumes less energy than a heavier vehicle. If the part is lighter by around 100 g, the part saves around 10 MJ in the life time of a vehicle if used in automobile vehicle. In aircraft use, the part having the same weight advantage saves around 10 GJ [36].

Considering an example of a part that is manufactured by CM and weighs 1.1 kg. If the part is redesigned and fabricated by AM so that it weighs 1 kg, AM contributes to make it lighter by 100 g. If the part is used in a vehicle, the vehicle becomes lighter by 100 g and saves maximum fuel energy, i.e., 10 MJ.

For fabricating this part, 1 kg of powder is consumed by the part while 39 kg of powder goes through some sort of deformation. If it is assumed that even 1 kg of powder is completely wasted, the part does not have any advantage of energy saving. Since producing 1 kg of powder requires 10 MJ while the part saves only 10 MJ in the whole lifetime of an automobile vehicle. Even if it is assumed that there are no other factors present which increase the energy requirement of production such as spillage of powder, extra consumption of beam energy during production, etc., AM does not have any benefit over CM for saving the energy.

AM will have benefit over CM in terms of energy saving if there is a high difference between the weight of the part made in CM and AM. It is possible if a bigger part is fabricated in AM [37] so that the difference between CM and AM part is not in grams but in kilograms, then AM helps save energy substantially. But if a bigger part is fabricated, its fabrication will itself consumes high energy since AM is an energy intensive process (refer Chap. 3). Thus if a bigger part is fabricated, both wasting and saving happen. Energy is wasted because the part is big while energy is saved because the part is going to be a life-long part of an automobile vehicle.

Energy will be saved if the following conditions are fulfilled: (1) AM makes a big part weighing several kilograms, (2) the part cannot be made in CM, (3) if the part is made in CM, the part is heavy, (4) AM makes the big part in such a way that the part is lighter by some kilograms when compared with CM, (5) AM consumes less energy for its fabrication, (6) AM does not consume energy indirectly such as spillage of materials, (7) the total energy fabrication in AM is less than in CM so that fabrication is sustainable, (8) the part is used in a vehicle, (9) the vehicle completes its lifespan.

The same weight difference, i.e., 100 g saves energy of 10 GJ when the part is used in aircraft rather than in automobile. In this case, even if a relatively large amount of powder is wasted during fabrication in AM, the part provides energy advantage. Moreover, there is no need to make a bigger part to gain energy advantage.

5.2   Material Wasting

If material is wasted in AM and the part is still cost-effective, there will be no incentive in manufacturing business to discontinue manufacturing of the part and help prevent material wastage. It will lead to an increase in carbon footprint and not help make environment sustainable. If AM machine is expensive, machine time is expensive. Hence, if material is not significantly expensive, the cost of the part is mainly governed by the machine time and not by the material cost. Consequently, the cost of the material lost during manufacturing is not high enough to increase the part cost. This is presently the case with metal AM machine where the cost of machine time is high in comparison to the cost of feedstock (steel or titanium). Hence, the loss of steel or titanium during manufacturing does not increase the cost of the part too high to discourage manufacturing of the part. If steel or titanium was replaced with gold, the loss of gold during manufacturing would increase the cost of the part high enough to discourage manufacturing of the part.

5.3   Cost

What if the cost of metal AM machine becomes low so that the cost of the material is not insignificant in comparison to the machine time. In this case, the cost of the part will be low, which will fluctuate with the cost of the material. If there is a material loss, the part cost will increase. Since the machine time cost is already low, the loss of the material during manufacturing will not make the part expensive. Hence, the loss of the material will not again be a reason to discourage manufacturing of the part.

5.4   Complexity

The cost to manufacture can be an overriding factor to take a decision whether to manufacture. But this cannot always be a factor. Considering an example where a part is manufactured because the part can be manufactured only in AM. This may be due to the part is complex or is consisting of many materials or has varied properties distributed within it. What if the fabrication of such parts in AM does not help conserve materials. It cannot be exactly estimated if the part is unique. But, it can be somewhat estimated. Considering an example of a complex part which is able to be manufactured in both AM and CM, but when the part becomes complex, it can be manufactured only in AM. Thus, it is not possible to exactly know for a complex part how much its fabrication in AM saves or wastes the material because there is no complex part in CM to compare with. Since there is a complex part available in CM which if used to compare with instead, the comparison can give a rough estimate.

If a part manufactured in AM is unique for which there is no precedence in CM and for which there is not even a near experience (as in case of a complex part) in CM, then there is not even a rough estimate to know whether the material is saved or wasted. In the absence of such knowledge, even it is felt that there is over-consumption of materials, it is difficult to term the over-consumption as wasting.

6   What Are the Optimum and Non-optimum Applications of Powder Bed Fusion?

Powder bed fusion (PBF) gives most complex metallic parts that can be achieved in AM.

6.1   Optimum Application

It has optimum applications such as making one-off part, emergency part, innovative part, and a batch of customized parts.

One-off Part

A part is made because it is cost-effective. Fabrication of one part does not necessarily imply that the part is expensive, but can be inexpensive if is fabricated by CM.

Emergency Part

A part is made because there is no other option to make it immediately. It is expensive but the cost is justified for emergency use.

Innovative Part

A part is made because other processes are unable to make such complex or multi-material part. Cost of the part can be high or low. The cost is incomparable because there is no other method to fabricate it alternatively.

Batch of Customized Parts

Though parts are neither for emergency use nor innovative but being in a batch make fabrication more resource-efficient than fabricating one-off part. Besides, making a batch by alternative methods such as machining or investment casting may not be efficient and convenient.

6.2 Non-optimum Application

PBF has non-optimum applications such as making a big part, many complex parts.

Big Part

Making a big part, which is neither resource-efficient nor energy-efficient such as making a big aircraft wing, which can better be made by CM is a non-optimum use of PBF. If fabrication by CM is expensive, AM will be a preferred choice (refer Chap. 5).

Many Complex Parts

Making hundred of those complex parts which can be fabricated by CM is a non-optimum use of PBF. When a complex part cannot be made by CM, its fabrication by AM is the only solution. If fabrication of such complex part is not resource-efficient and is not sustainable in the long term, still the fabrication is justified because it is one-off part. But if hundred of such complex parts are made, it is a burden to the environment. Moreover, if that part is cost-effective, the fabrication goes unabated which in turn creates more problems to the environment.

References

1. Hopkins, N., Jiang, L., & Brooks, H. (2021). Energy consumption of common desktop additive manufacturing technologies. *Clean Engineering Technology, 2*, 100068.
2. Durão, L. F. C. S., Christ, A., Zancul, E., et al. (2017). Additive manufacturing scenarios for distributed production of spare parts. *International Journal of Advanced Manufacturing Technology, 93*, 869–880.
3. Hanssen, J., Moe, Z. H., Tan, D., & Chien, O. Y. (2013). Rapid prototyping in manufacturing. In *Handbook of manufacturing engineering and technology*. New York: Springer. https://doi.org/10.1007/978-1-4471-4976-7_37-2.
4. Khajavi, S. H., Partanen, J., & Holmström, J. (2014). Additive manufacturing in the spare parts supply chain. *Computers in Industry, 65*(1), 50–63.

5. Culmone, C., Smit, G., & Breedveld, P. (2019). Additive manufacturing of medical instruments: A state-of-the-art review. *Additive Manufacturing, 27*, 461–473.
6. Petersen, E. E., Kidd, R. W., & Pearce, J. M. (2017). Impact of DIY home manufacturing with 3D printing on the toy and game market. *Technology, 5*(3), 45.
7. Colorado, H. A., Mendoza, D. E., & Valencia, F. L. (2021). A combined strategy of additive manufacturing to support multidisciplinary education in arts, biology, and engineering. *Journal of Science Education and Technology, 30*, 58–73.
8. Huang, Y., Leu, M. C., Majumder, J., & Donmez, A. (2015). Additive manufacturing: Current state, future potential, gaps and needs, and recommendations. *Journal of Manufacturing Science and Engineering, 137*(1), 014001.
9. Prabhu, R., Miller, SR., Simpson, TW., & Meisel, NA. (2018). Teaching design freedom: Exploring the effects of design for additive manufacturing education on the cognitive components of students' creativity. In *Proceedings of ASME 2018 International Conference*, Quebec, Canada, August 26–29.
10. Calhoun, S. J., & Harvey, P. S., Jr. (2018). Enhancing the teaching of seismic isolation using additive manufacturing. *Engineering Structures, 167*, 494–503.
11. Anastasiadou, C., & Vettese, S. (2021). Souvenir authenticity in the additive manufacturing age. *Annals of Tourism Research, 89*, 103188.
12. Thurn, L. K., Balc, N., Gebhardt, A., & Kessler, J. (2017). Education packed in technology to promote innovations: Teaching based on additive manufacturing on a rolling lab. *MATEC Web Conference, 137*, 02013.
13. Lim, S., Buswell, R. A., Le, T. T., et al. (2012). Developments in construction-scale additive manufacturing processes. *Automation in Construction, 21*, 262–268.
14. Ferreira, T., Almeida, HA., Bártolo, PJ., & Campbell, I. (2012). Additive manufacturing in jewellery design. In *Proceedings of the ASME 2012 11th Biennial Conf Nantes*, France, pp. 187–194.
15. Lipton, J. I., Cutler, M., Nigl, F., et al. (2015). Additive manufacturing for the food industry. *Trends in Food Science and Technology, 43*(1), 114–123.
16. Labeaga-Martínez, N., Sanjurjo-Rivo, M., Díaz-Álvarez, J., & Martínez-Frías, J. (2017). Additive manufacturing for a moon village. *Procedia Manufacturing, 13*, 794–801.
17. Snyder, MP., Dunn, JJ., & Gonzalez, EJ. (2013). Effects of microgravity on extrusion based additive manufacturing. In *Space Exploration Conference*, Sep 10–12, San Diego, CA.
18. Boer, J., Lambrechts, W., & Krikke, H. (2020). Additive manufacturing in military and humanitarian missions: Advantages and challenges in the spare parts supply chain. *Journal of Cleaner Production, 257*, 120301.
19. Wilson, J. M., Piya, C., Shin, Y. C., et al. (2014). Remanufacturing of turbine blades by laser direct deposition with its energy and environmental impact analysis. *Journal of Cleaner Production, 80*, 170–178.
20. Petrat, T., Graf, B., Gumenyuk, A., & Rethmeier, M. (2016). Laser metal deposition as repair Technology for a gas turbine burner made of inconel 718. *Physics Procedia, 83*, 761–768.
21. Li, L., Li, C., Tang, Y., & Du, Y. (2017). An integrated approach of reverse engineering aided remanufacturing process for worn components. *Robotics and Computer-Integrated Manufacturing, 48*, 39–50.
22. Chandra, S., Tan, X., Wang, C. et al. (2018). Additive manufacturing of a single crystal nickel-based superalloy using selective electron beam melting. In *Proceedings 3rd International Conference Progress Addit Manuf (Pro-AM)* (pp. 427–432).
23. Basak, A., Acharya, R., & Das, S. (2016). Additive manufacturing of single-crystal Superalloy CMSX-4 through scanning laser epitaxy: Computational modeling, experimental process development, and process parameter optimization. *Metallurgical and Materials Transactions A: Physical Metallurgy and Materials Science, 47*, 3845–3859.
24. Shinde, M. S., & Ashtankar, K. M. (2017). Additive manufacturing-assisted conformal cooling channels in mold manufacturing processes. *Advances in Mechanical Engineering, 9*, 5.

25. Diegel, O., Schutte, J., Ferreira, A., & Chan, Y. L. (2020). Design for additive manufacturing process for a lightweight hydraulic manifold. *Additive Manufacturing, 36*, 101446.
26. Feng, S., Kamat, A. M., & Pei, Y. (2021). Design and fabrication of conformal cooling channels in molds: Review and progress updates. *International Journal of Heat and Mass Transfer, 171*, 121082.
27. Kuo, C. C., Jiang, Z. F., & Lee, J. H. (2019). Effects of cooling time of molded parts on rapid injection molds with different layouts and surface roughness of conformal cooling channels. *International Journal of Advanced Manufacturing Technology, 103*, 2169–2182.
28. Scudino, S., Unterdörfer, C., Prashanth, K. G., et al. (2015). Additive manufacturing of Cu–10Sn bronze. *Materials Letters, 156*, 202–204.
29. Brøtan, V., Berg, O. A., & Sørby, K. (2016). Additive manufacturing for enhanced performance of molds. *Procedia CIRP, 54*, 186–190.
30. Mitra, S., Castro, A. R. D., & El Mansori, M. (2019). On the rapid manufacturing process of functional 3D printed sand molds. *Journal of Manufacturing Processes, 42*, 202–212.
31. Hawaldar, N., & Zhang, J. (2018). A comparative study of fabrication of sand casting mold using additive manufacturing and conventional process. *International Journal of Advanced Manufacturing Technology, 97*, 1037–1045.
32. Druschitz, A., Williams, C., Snelling, D., & Seals, M. (2014). Additive manufacturing supports the production of complex castings. In M. Tiryakioğlu, J. Campbell, & G. Byczynski (Eds.), *Shape casting: 5th Int Sym* (pp. 51–57). Cham: Springer.
33. Carneiro, V. H., Rawson, S. D., Puga, H., et al. (2020). Additive manufacturing assisted investment casting: A low-cost method to fabricate periodic metallic cellular lattices. *Additive Manufacturing, 33*, 101085.
34. Kruzhanov, V., & Arnhold, V. (2012). Energy consumption in powder metallurgical manufacturing. *Powder Metallurgy, 55*(1), 14–21.
35. Cordova, L., Campos, M., & Tinga, T. (2019). Revealing the effects of powder reuse for selective laser melting by powder characterization. *JOM, 71*, 1062–1072.
36. Hettesheimer, T., Hirzel, S., & Roß, H. B. (2018). Energy savings through additive manufacturing: An analysis of selective laser sintering for automotive and aircraft components. *Energy Efficiency, 11*, 1227–1245.
37. Hassen, A. A., Noakes, M., Nandwana, P., et al. (2020). Scaling up metal additive manufacturing process to fabricate molds for composite manufacturing. *Additive Manufacturing, 32*, 101093.

Chapter 7
Fabrication Strategy

1 What Gives Creativity in AM?

Following gives creativity in AM: support structure, mixing of material, design, changing parameter.

1.1 Support Structure

Creating a new product requires systematic study, which involves simulation, experimentation, trial and error methods or a combination thereof, etc. [1]—these are the best methods but at the same time are the reason for delay for not trying to make new products frequently. Fabrication of support structure (SS) is not always a task as serious as making an end-use product because SS is ultimately removed and will not be the life-long part of the product. But the reason that makes SS unimportant is also the reason why it is important.

Making SS gives an opportunity to check whether it can also be part of a final product or to check whether there is a significant change in the quality of mechanical properties as introduction of SS can help transfer heat fast [2]. Introduction of SS can give extra strength to overhang and retaining the SS afterwards can be the best option because SS is lighter in weight. If SS needs to be removed, why it should be removed immediately. It can be checked whether it has any use afterwards, it can be a fixture as well for carrying out machining afterwards [3].

S. Kumar, *Additive Manufacturing Solutions*,
https://doi.org/10.1007/978-3-030-80783-2_7

1.2   Mixing of Material

AM is for adding materials and made by adding materials. Thus any material, in principle, can be added unless forbidden by law. This gives opportunities to check what happens if some unknown materials can be mixed, or if some component of a mixture of materials can be decreased and increased. For example, adding nanoparticles has shown to improve the processability of metallic alloys and helped make crack-free parts [4] (refer Chap. 2).

Mixing of materials is not new, it is as old as the history of materials. They are mixed frequently in dies and molds and new products are formed. But, what is new is new achievable properties unachieved in the past plus lack of predictability of properties. Presence of two opposite realities gives rise to creativity.

1.3   Design

Earlier, it was a piece of paper to draw a design on and explore various possibilities. The piece of paper provided the safety that the failure to draw the right design would not cost a fortune. AM has extended the safety net to three dimension, the failure of a 3D physical model will not henceforth cost a fortune. The hesitation not to 3d-print a design because the design is not yet perfect is already gone.

A guiding principle that manufacturability of a design should be known while making the design is no longer strict, if not obsolete. Limitations to make models due to traditional tools such as lathe machine, laser cutter, bandsaw have gone as well [5]. Henceforth, a number of 3D physical models can be made conveniently, which has given rise to creativity [6, 7].

1.4   Changing Parameter

Combination of various experimental parameters gives various results. For examples, varying some parameters can give rise to porous products having various degrees of porosity leading to many applications, while varying other parameters can give rise to various values of strength suitable for different applications. Thus changing parameters different products for different applications can be created (refer Chap. 4).

In photopolymer based process, by changing parameters different amount of curing can be realized leading to different properties in a product. If these semi-cured parts can further be post-treated in a curing oven, again a range of properties can be obtained. Hence, a parametric study for different applications leads to different results that will be suitable for some other products not envisaged during investigation.

What if every single powder out of thousand powders flowing from a nozzle and striking a substrate is important not because every powder should not be wasted but because any powder can create an irreparable defect. In this dire situation, is it possible to overcome the deeds and misdeeds of each powder. For making a single crystal, a single powder coming from the nozzle can be a possible nucleation site for the growth of a crystal. If there are more than one such powder, the development of a single crystal is an impossibility—when a single pore or single crack counts, when the orientation of the growth counts, when conflict between two growth orientations restricts the height of the single crystal formed. In spite of these problems, optimization of parameters leads to a point when the crack gets self-repaired and nucleation sites get remelted to form a single crystal [8, 9] (refer Chap. 6).

2   What Is Support Structure in AM?

If a part, which consists of two parallel walls separated by a distance and connected by a plate on top of them, is to be made using AM, a support structure (SS) is required to make the part (Fig. 7.1a).

Considering a liquid deposition process (DP) where drops from a nozzle are a means to add material. For making walls, the nozzle will add liquid on a base plate (substrate) to make the first layer that will then become a base on which further drops will be deposited, the wall will thus start taking form. The ongoing deposition will not collapse because there is always some base available, sometimes it is a base plate, some other times it is the first layer or the second layer and so on, to let the drop rest and have a chance to become solid. If there was no base available, the drop would fall on the ground. The drop will not hang on air on its own even if an important structure is needed at that height.

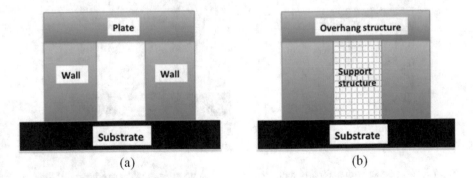

Fig. 7.1 (a) Design of a part which requires SS in liquid DP, (b) application of SS

2.1   Need for Support Structure

Thus making both walls of the part is easier. During their fabrication neither of the walls is giving drops any chance to fall and then not to contribute to the construction of the walls. What will happen if the nozzle moves from one wall toward another wall to construct a plate that will connect both walls. Liquid drops from the nozzle will fall between two walls. Thus AM will not be able to make even such a simple part. There are some ways to solve the problem.

The first way is to create a false plate underneath the plate so that the drop will have a place to be at. This false plate is supporting the drop so that an actual plate will be formed. This false plate is called SS (Fig. 7.1b). It is created by fabricating a number of other walls and horizontal section between these two walls. These supporting walls and section may not be as solid and strong as two main walls, therefore their fabrication is easier. These supporting walls stand on the same level from where two main walls are standing. Hence, if a tall part is constructed, fabricating a tall SS is an enormous task (Fig. 7.2b). The job of SS is done when the plate is formed, SS is having no purpose thereafter.

Figure 7.2a shows two overhangs to be fabricated. For making an upper overhang, SS from lower overhang can work. In the absence of lower overhang, longer support is required starting from the substrate. Figure 7.2b shows for fabricating extended upper overhang, support from the lower overhang is not sufficient. If making a tall SS is difficult, fabricating the part shown in Fig. 7.2b is not possible while that shown in Fig. 7.2a can be fabricated. If removing the SS causes damage on the surface of the part, the part shown in Fig. 7.2a cannot be fabricated.

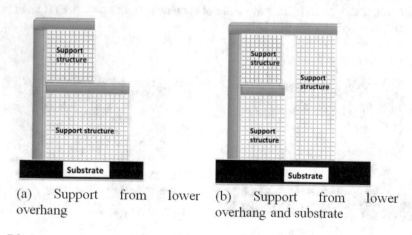

(a)   Support   from   lower
overhang

(b)   Support   from   lower
overhang and substrate

Fig. 7.2 Support structures for different overhangs. (**a**) Support from lower overhang. (**b**) Support from lower overhang and substrate

2.2   *Changing the Orientation*

The second way is to find a method so that there will never come a chance when drop will not be having a base or layer as it had when the wall was getting constructed. If the plate is made first and the walls are constructed on it thereafter, there is no chance that any section of the part is not completed because drops fall away. Thus by changing orientation of the design, the same part can be constructed without the help of SS (Fig. 7.3a). But, changing the orientation is not possible in all types of design. If a design shown in Fig. 7.3b is oriented at 90° or 180° or 270° from the build direction (z-axis), there is always an overhang needing SS.

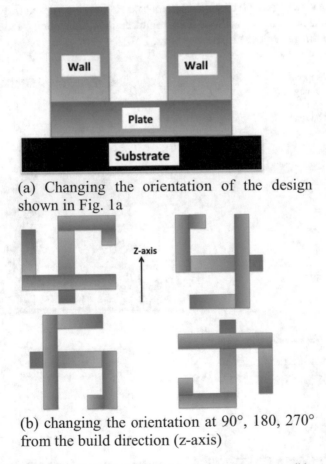

(a) Changing the orientation of the design shown in Fig. 1a

(b) changing the orientation at 90°, 180, 270° from the build direction (z-axis)

Fig. 7.3 Changing the orientation to avoid SS (**a**) is possible, (**b**) is not possible

2.3 Self-Support Structure

There is a third way if design is somewhat different. If the plate of the part is not horizontal and planar but is having a curve similar to an arch, depending upon the curve there is no need of either SS or changing orientation (Fig. 7.4). The drop from the nozzle will hang on the wall and will not be detached due to the gravitational force. This partial attachment of the drop, if continues for every drops, leads to an arch type structure. Depending upon the thickness, length, and angle, all types of arch structures are not possible. The part thus formed is not having same property if the same part is formed with the help of SS. The third way thus can be utilized when both changing the orientation and creating SS is not possible.

In extrusion based process, if there is a continuous fiber reinforcement [10], deposited bead does not fall through gap, thus no support is required (Fig. 7.5a). But for thermoplastic filament without fiber reinforcement, deposited bead may either fall through the gap or cave in the gap (Fig. 7.5b).

Fig. 7.4 Self-support structure—overhang structures for which support is not be required

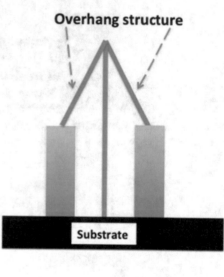

Overhang structure

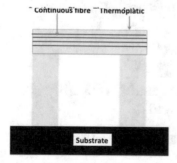
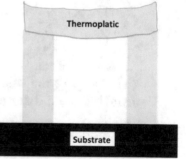

(a) Non-bending of deposited bead (b) Bending of deposited bead

Fig. 7.5 (**a**) No SS required in continuous fiber reinforced extruded bead, (**b**) bending of non-reinforced thermoplastic extruded beam in the absence of SS

2.4   Four D's

Whether extruded material or extrudate falls through the gap (Fig. 7.5b) or not depends mainly on four D's: day, diameter, density, and distance.

The extrudate will never fall if it cools fast enough to become stiff before it starts sagging. How fast it cools depends upon which type of day it is, on the day of extrusion. If day is warm, it will take time to cool and has chance to sag. If day is cold, it will cool fast and has chance not to sag.

Besides day, it depends on how thick it is or what is its diameter. If it has large diameter, it will have small surface-to-volume ratio and it will take time to cool and has chance to sag. If it has small diameter, it will have high surface-to-volume ratio and it will cool fast.

Even it has small diameter—it does not matter if it is made up of dense material. If it has high density, it will have large weight, and it needs to cool completely in order to save itself from sagging. If it partially cools, its density will dominate and because of relatively high density it will fall through the gap.

Even it has high density and large diameter—it does not matter if the distance of the gap is small. So that before it starts to fall, it has already travelled the distance and crossed the gap. On the contrary, if the distance of the gap is large, it may need support.

2.5   Reusable Support

There is another way—instead of using false plate, real (reusable) plate will be used as SS so that there will be no need to make SS, there will also be no need to recycle SS. The real plate can again be used for the next fabrication. Where overhang is required, a metallic plate is elevated from the platform to support the fabrication of the overhang. After the fabrication, the plate is detached from the overhang for further use. Thus, if platform is modified to include the provision (nut, bolt, fixture, jack) for using the real plate, the need for making the false plate is fulfilled [11] (Fig. 7.6).

2.6   Bed Process

Creating a part in DP means ensuring that no bit of its material falls through any gap because there is already a provision to fill the gap by making SS. But what if no such gap exists. What if the process is such that it does not allow any material to fall because the process never created a gap to fall through. This happens in bed process (BP).

In BP, it is the bed which acts as SS. If bed is made from powder, the powder is acting as SS. In the case of slurry or photopolymer bed, it is the slurry or

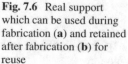

Fig. 7.6 Real support which can be used during fabrication (**a**) and retained after fabrication (**b**) for reuse

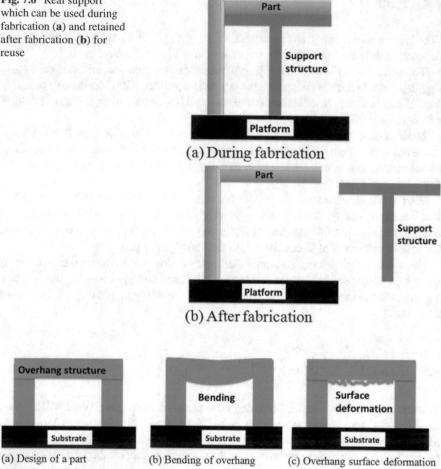

(a) During fabrication

(b) After fabrication

(a) Design of a part   (b) Bending of overhang   (c) Overhang surface deformation

Fig. 7.7 Deformation of (**a**) an overhang structure resulting in (**b**) bending of overhang and (**c**) higher underneath surface roughness of overhang

photopolymer acting as SS. The advantage of bed being SS is that no part fails to be fabricated in BP because the design of the part cannot accommodate SS that cannot be accommodated when is tried in DP. Having bed as an all-prevailing support facilitates BP to make parts complex than that possible in DP (refer Chap. 2).

BP does not pose the same difficulty that arises in DP while creating SS. But being free from such difficulty does not mean being free from all difficulties. BP comes up with its own set of difficulties when a robust support is needed for making a part. Powder in powder bed process needs to provide that support but powder or powder layer is not a solid structure similar to that formed in DP. A powder layer will not able to act as a solid plate. Though powder as SS is better than having no powder to support but powder layer will cave in when a heavy section (Fig. 7.7a) of the part is built over it (Fig. 7.7b). If caving in does not let the section to deform or bend, at least it will not allow to have smooth underneath surface (Fig. 7.7c).

Powder is having different thermal properties than a solidified layer made from the same powder. A section made on a solidified layer acquires different properties than when made on a powder layer [12]. The section made on a powder layer warps upward giving rise to curl surface rather than a straight surface [13] while the section made on a solid support is free from such warping [14]. Therefore, fabrication on a powder layer which is acting as SS requires extra optimization in order to have high quality of downfacing surface (Fig. 7.7c). This extra optimization is again having limitation. To avoid this situation, reliance on a powder layer support is minimized and SS is fabricated for a complex part which is having a number of downfacing surfaces. If an overhang is built at certain orientation from the build direction (Fig. 7.4), SS is not required and the quality of downfacing surfaces can be improved [13].

In photopolymer bed process, it is the photopolymer liquid which acts as SS. But, making a section on liquid is different from making a section on solidified and cured photopolymer. During solidifying a certain thickness on a liquid bed, the depth of curing needs to be maintained such that curing does not trespass and solidify the liquid which is acting as a support underneath. The trespassing causes a deterioration in surface and dimension. If a section is made on a solidified layer, exceeding the curing depth does not cause a change in surface and dimension but is accommodated as overcuring of the underlying solidified layer. Thus making SS instead of taking help of liquid as SS will facilitate maintaining quality of that type of parts whose quality happens to deteriorate due to the liquid support [15].

2.7 Various Types

Selection of SS depends upon whether it will be:

- Same or different material [16],
- Porous or non-porous,
- Equally strong or weaker [17],
- Lattice or solid type [18],
- Easily removable [19].

If the material of SS is different from that of a main structure, there is possibility that during fabrication and removal of SS both materials contaminate if they mix. Contamination causes these materials not to be recycled or reused. To avoid contamination, it is better that the support and main structure are made from the same material (refer Chap. 2).

SS is made from a weak material since SS has no use after the fabrication of a part. Since SS does not become a section of the final part, the possibility that the strength SS will contribute to the final part does not exist. A weak material is less expensive than the material a main structure. Thus using weak material does not let the fabrication of the complex part become expensive when the complex part requires a number of SS.

An ideal SS should be weak enough so that it can be easily broken and removed. Fabricating SS to be weaker is easier than fabricating a main structure to be strong. Porosity causes weakness in a part and removing the porosity requires finding the right parameters and fabrication conditions. If porosity does not need to be removed, there is no such fabrication requirement. Consequently, if SS needs to be fabricated which is not expected to be strong, its fabrication will be easier.

For example, for making SS, scanning can go fast while for fabricating the main structure scanning does not go fast so that there should not be any area left partially scanned causing a gap or pores. Scanning an area can again go faster for making SS if the overlap between two consecutive scans is not as much as in the case of making the main structure. For making SS unlike the main structure, there is no need to bring two consecutive scans closer and increase the overlap if there is no need to decrease porosity. If two consecutive scans are kept far apart so that the gap between them is far more, resulting in not a zero overlap but a negative overlap, it will increase fabrication speed of SS. Thus fabricating SS can be faster than fabricating the main structure unless fabrication is not hindered by variation in geometry of the structures.

Such gap will cause porosity and bring unknown property in SS. But SS is not required to have a well-defined property as the main structure should have. Any property of SS is acceptable as long as it is able to do its only duty that is supporting the main structure. Such gap between two consecutive scans, if taken to higher level will help make a lattice structure. Though, fabrication of a lattice structure is different in different AM processes and increasing the gap between two consecutive scans is not the only requirement. Nevertheless, fabrication of a lattice structure means saving the energy, time, and material for fabrication. SS thus made of lattice type will facilitate faster fabrication of SS. Since lattice structure consumes less material, less material will be finally wasted if there is no plan to recycle SS material.

To remove SS well, it can be made from water-soluble material such as sugar or polyvinyl alcohol so that after the fabrication of a polymer part, SS is removed by dipping the part in water [20]. But this type of SS material is not amenable to be fabricated by all AM processes. In bed system, introducing another material to create weak SS requires modification of the system [21].

Though there are many designs that cannot be fabricated in AM because they cannot be fabricated without SS, there are also many designs that cannot be fabricated in AM because they cannot be fabricated with SS (e.g., there is no space for SS to be included, if it is included it cannot be removed).

3   What Can Be a Product Fabrication Strategy in AM?

3.1   Need for Strategy

AM processes are not yet developed to make all complex products. The process will always remain undeveloped for a certain complex product. Therefore relying on developing only the process will not always help, there requires a strategy to maximize the benefit from an existing process.

3.2   Problem with AM System

AM builds an object either on a platform or on a substrate (Fig. 7.8). After fabrication, the object is removed from the platform or the substrate (Fig. 7.8c). Figure 7.8a shows a schematic diagram of a deposition process (DP), Fig. 7.8b shows an object built on a substrate and, Fig. 7.8c shows the object without the substrate. Fig. 7.8d, e show schematic diagrams of bed system (BS) using platform and substrate, respectively.

The use of the substrate is to help fabricate an object, its use is over after the removal of the object. The substrate has no contribution to the fabrication or shaping except that it allows the object to be built over it. AM does everything to create a shape. Since AM has limitation, the shape that is created shows the limitation.

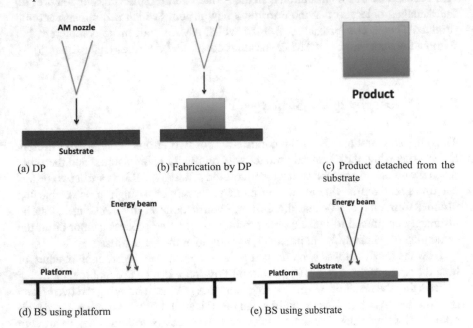

Fig. 7.8 Deposition system (DS) (**a, b, c**) and bed system (**d, e**)

If instead of a plane rectangular substrate, various shapes of the substrate are used, AM will be free from doing all shaping and will make only those features which the substrate does not contain [22]. Thus the product that is formed is an improved substrate.

3.3 What Should Be the Strategy

For fabricating a product, one process will be used to make a pre-product that in the form of a substrate will be fitted in an AM system to convert it into a product. Therefore, one process does what it can do its best to contribute for making a product and AM finalizes by doing what is left to do [22] or vice versa [23]. Alternatively, one AM process does what it can do its best to contribute for making a product and another AM finalizes by doing what is left to do [24, 25].

For example, for fabricating a mold, a block can be machined and some cavities can be milled on it while some features on the cavity will be added by laser solid DP. Thus the use of machining and milling process saves AM from wasting its resources on a task which can be done conveniently and economically by another process (machining) [26].

Another example, which is similar to the concept but not exactly same, is the remanufacturing of a damaged object. In this example, using AM, the damaged object is repaired or refurbished. Thus AM does not fabricate the whole object. But AM contributes. The contribution is in the form of adding some missing features on the damaged object so that there forms a new product which is a remanufactured product. Thus a new product is formed which AM did not make completely, but without the contribution of AM the product could not be formed (refer Chap. 6).

3.4 Philosophy of the Strategy

Thus, the new product shows the contribution of two efforts done at two different times—one when the damaged product was originally manufactured and the other when the damaged product was remanufactured. These two efforts are at two different times and can be due to two different processes or a single process, but the product formation shows that the efforts, even if done at different times, can be ultimately combined to make a new product, which one process cannot do at the same time or which different processes cannot do at the same time.

How the result of these two efforts separated by time can be brought together, or how the result of these two efforts separated by space can be brought together.

Bringing the result of two efforts together does not mean bringing the two efforts or processes together. This does not mean combination of two processes and development of a new process or optimization of two processes to find better result with that combination and optimization. Moreover, this does not mean adjusting one process in the light of other process and to deal with the compromise of two processes and accept the shortcoming of one process due to the presence of another process. This does not mean to accept the lesser solution because two processes are not compatible. This does not imply to fail not because two processes are not compatible but their systems are not developed to be compatible.

This means to be benefitted from two processes without letting one process interfere in the business of other process. This implies to be benefitted from the same process without letting the process (done at one time) interfere in the business of the same process (done at different time) because the same process is used at two different times. The same process is used at two different times because the same process is not able to make same feature twice on the same product when two features are fabricated at the same time (Figs. 7.12, 7.13).

Bringing the result of two efforts together does not mean making a pre-product by one process in the presence of other process and let the pre-product be converted to a product by other process. Bringing the result of two efforts together only means making a pre-product by one process in the knowledge of other process but in the absence of other process. So that the conversion of the pre-product to the product can happen because only those pre-products are fabricated which have convertibility with respect to the other processes.

Hence, a pre-product which is converted to a product using laser solid DP has convertibility with respect to this process but does not have convertibility with respect to powder bed fusion (PBF). For example, a pre-product can be converted to a mold having cooling channel using PBF but not using laser solid DP. This brings limitation to the pre-product that can be fabricated and therefore the product that can be fabricated. This brings limitation to the pre-product that can be fabricated with respect to a process but this does not bring limitation with respect to other process. Thus, what is limitation with one process is not necessarily the limitation with other process.

These limitations show the limitation of a process when a pre-product is brought in a system where this pre-product will be converted to a product by the process. But this limitation of the process is less than the limitation of the same process when the process (AM) will be combined with the other process (for example, machining) and the process will be forced to be compatible with the other process (e.g., machining).

When a pre-product is brought to a process and a product is formed—there is a limitation of the process. When pre-product is not brought and the whole product is formed by a process—there is more limitation of the process. When a pre-product is not brought and the whole product is formed by a combination of the processes—there is more limitation of each process. Consequently, there is always a limitation of the process—sometimes the limitation is more and sometimes the limitation is less.

This gives an inference that a process needs to be developed for getting a product to be formed. More development of the process, better the product it is. A developed process is better than an undeveloped process because the developed process gives a better product while the undeveloped process does not able to give such better product. Though an undeveloped process is not able to give a better product, but the developed process is not able to give a better product as well for which the developed process is not developed enough. There comes a point when both developed and undeveloped processes are not able to give a product when the requirement of the product becomes higher.

Then what is going to happen, the development of a process is an open-end process, it will just go on. The solution is not the continuous development of a process but a manufacturing strategy. It does not mean that manufacturing strategy is better than the development of a process, it only means that development of a process without manufacturing strategy is not better.

3.5   Problem and Solution

Problem

A product as shown in Fig. 7.9 needs to be fabricated with the help of PBF and machining. If the whole product is made by PBF, it is expensive and takes considerable time. On the other hand, if the whole product is made by machining, machining has to remove materials from a block having bigger dimension than the product. Thus a lot of materials need to be removed so that thin features 1, 2, 3, 4 will be left after machining. These thin features have propensity to break due to machining and if any of the features is broken, machining needs to be repeated from the start. Therefore the product needs to be fabricated by choosing both processes (PBF and machining) optimally.

Solution

PBF was tried to make this product. Figure 7.9 shows the product is made in the same orientation in which the design (Fig. 7.10) is shown. In this orientation, two features 3, 4 showed the overhang problem and their downfaces show inaccuracies in the dimension. Thus the fabrication of the product failed.

The product was tried in the other orientation possible (Fig. 7.11), but it again showed the same overhang problem but on different features 1, 2 and downfaces of the cavity. Thus the fabrication of the product failed again. There is no more orientation left in which the product can be fabricated without having overhang problems.

Failure in the fabrication shown in Figs. 7.10, 7.11 demonstrates that PBF failed in fabricating the product though all possible orientations were tried. Thus the process is not developed enough to make this product. But Figs. 7.10, 7.11 show the process succeeded partially. Figure 7.10 shows it succeeded in making all other features except features 3, 4 while Fig. 7.11 shows it succeeded in making features 3, 4. Though the process is not successful, a strategy can be successful. The process

Fig. 7.9 A product having four features named 1, 2, 3, 4 and a rectangular cavity

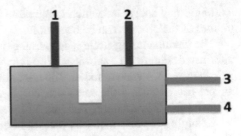

Fig. 7.10 Product fabricated by PBF in one orientation showing problems in features 3, 4

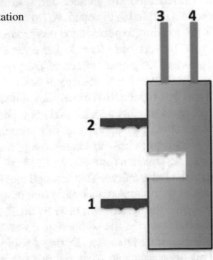

Fig. 7.11 Product fabricated by PBF in other orientation showing problems in features 1, 2 and cavity

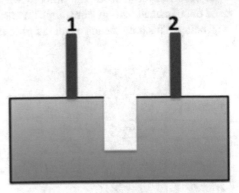

has already shown partial successes—one partial success at one orientation and the other partial success at the other orientation.

If a strategy comes which captures the partial success happened in one orientation and the partial success happened at the other orientation, if a strategy is there to capture the partial success at one time and the partial success at the other time, the product fabrication will be successful. If there is a manufacturing strategy to combine two successes which happened (though separated by time and attempt), failure of the process can be overcome.

The manufacturing strategy is to make only those parts of the product that can be fabricated. Figure 7.12 shows a part made by PBF in which fabrication of problematic features 3, 4 was excluded. The next attempt was done to work on the same part or pre-product using the same process, i.e., PBF but by changing the orientation (Fig. 7.13a), the remaining features 3, 4 were built on it afterwards (Fig. 7.13b).

Changing the orientation is possible because both upper and lower surfaces in this orientation are parallel and normal to each other. Therefore when the pre-product is fitted in the platform (to remain under the powder bed plane), extra effort is not required to position the pre-product in order to make its upper surface parallel to the powder bed plane. In the absence of such parallelism between powder bed plane and the upper surface of the pre-product, either due to the geometry of the pre-product or due to the lack of positioning, it will not be possible to make features 3 and 4 of the required dimension and shape.

Thus a product which could not be built by a process was built by the same process but with the application of a strategy. The process was tried twice, the system was utilized twice—this is a strategy that worked to beat a process using the same process. This is a triumph of a strategy over the process. But this triumph is not a triumph but a failure. This triumph can be a technical triumph but this triumph is a failure from the sustainability point of view or from the management point of view.

When the system was used twice, it consumed energy twice, it wasted material twice. Moreover, the whole object was fabricated by a single process without considering other processes. During finding the strategy, bigger picture was forgotten. This does not mean when the process was developed enough and the system was used once instead, energy consumption and material wastage were occurred once, it was better. This only means when the process was developed enough and the system

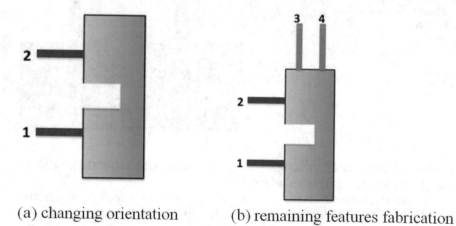

(a) changing orientation   (b) remaining features fabrication

Fig. 7.12 Pre-product fabricated by PBF in which fabrication of features 3, 4 was not tried

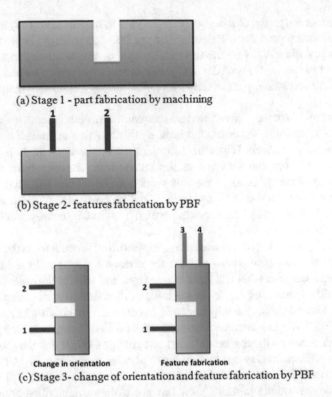

(a) Stage 1 - part fabrication by machining

(b) Stage 2- features fabrication by PBF

Change in orientation   Feature fabrication

(c) Stage 3- change of orientation and feature fabrication by PBF

Fig. 7.13 Product fabricated by PBF by changing the orientation of part or pre-product fabricated by PBF

was used once instead, energy consumption and material wastage were occurred once, it was also not better but it was worse.

There are two cases. In the former case, a process is not developed, a strategy is made and bigger picture is forgotten. In the latter case, a process is developed, therefore no strategy is made and bigger picture is forgotten. Either of the two cases is not appreciable. Because, in the former case, just a technical strategy was made while in the latter case, there is lack of acknowledgement of a strategy.

Fabricating the whole object (shown in Fig. 7.9) by a single process (PBF) is itself a wrong strategy irrespective of whether the process is used twice or can be used once. The design (Fig. 7.9) shows that it consists of two types of sections—one type consists of fine features 1, 2, 3, 4, other type does not consist of fine features.

The design can thus be divisioned into two sections—one having fine features requiring a process which is better than other processes to make it, and the other having remaining section requiring any process. If it is any process, a process which is fast, economic, and convenient can be a process of choice. Thus the design which is divisioned into two sections requires a division of labor—one which is advanced (PBF) will make fine features and which is economic (machining) will do the bulk work.

Thus the investigation of a design (Fig. 7.9) led to the allocation of task to different manufacturing processes. This allocation of task is a strategy. This allocation of task is reached after knowing the technical merits of two manufacturing processes with respect to two sections of the design. Thus this strategy is a technical strategy. But, this technical strategy is not only a technical strategy, it is more than a technical strategy.

This technical strategy takes into the account the economic consideration when it considers machining for doing the bulk work, thus this technical strategy is an economic strategy as well. When this technical strategy assigns the bulk work to be done by machining, the strategy at the same time snatches bulk work from PBF. When the strategy deprives the bulk work from being done by PBF, it saves the environment from more carbon emission, it makes environment more sustainable. Thus this technical strategy becomes more than a technical strategy when it does its bit to save the environment.

Thus just by selecting machining as a manufacturing process to do the bulk work, the strategy wins in both front—one on the economic front and the other on the environment front. But what if it loses at one front and wins at the other front. What if selecting the machining for doing the bulk work is environment-friendly but not economic. The machining is not economic because the machining has not become expensive but PBF has become inexpensive. The PBF becoming inexpensive or expensive depends on the market and does not change the fact that the production of powders as well as making a part with those powders requires energy. In this case, the selection of PBF instead of machining for doing the bulk work will be a better strategy from economic point of view but not from sustainability point of view. Again, the selection of machining instead will be a better strategy from sustainability point of view but not from economic point of view (refer Chap. 6).

This brings a question—which strategy is better: PBF + machining or PBF + PBF, not from sustainability point of view and not from economic point of view but from agility point of view, assuming that both strategies make equally satisfactory products.

PBF + machining is a better strategy than PBF + PBF because PBF + machining is faster than PBF + PBF due to PBF being a slower process. Since PBF + PBF does not mean working in a PBF system once but twice (Figs. 7.12, 7.13), thus this strategy does not gain time advantage because the system does not make product in one go—the system still needs to shut down to change the orientation of the pre-product (Fig. 7.13). If the meaning of PBF + PBF was to use PBF once, PBF + PBF would have some more time to outrace PBF + machining.

Figure 7.14a shows the bulk part of the design (Fig. 7.9) which is fabricated by machining. On this part or pre-product, two features 1,2 are built using PBF (Fig. 7.14b). This part needs to be again used in PBF by changing the orientation (Fig. 7.14c). Using a strategy PBF + machining, the part is built in three stages.

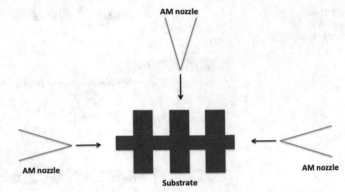

(a) A pre-product in the form of substrate is fitted in DS

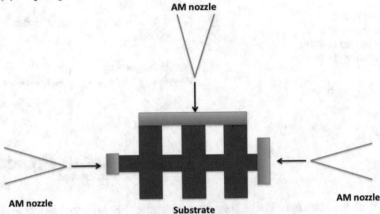

(b) Fabrication of various features on the substrate

(c) Final product is subtrate plus features

Fig. 7.14 Three stages in product fabrication by machining and PBF. (**a**) Stage 1—part fabrication by machining. (**b**) Stage 2—features fabrication by PBF. (**c**) Stage 3—change of orientation and feature fabrication by PBF

3.6 Compatibility Between Processes

Figure 7.15a shows a complex substrate fitted in a system equipped with three nozzles to deposit on the substrate. These three nozzles have three build directions and are acting on three different surfaces so that materials can be deposited on three

Fig. 7.15 Deposition
based combined AM: (a)
process, (b) fabrication, (c)
product

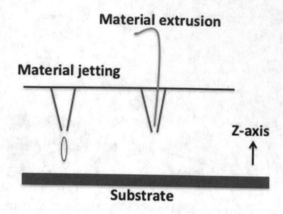

Fig. 7.16 Two DP:
material jetting and
material extrusion in one
build direction (z-axis)

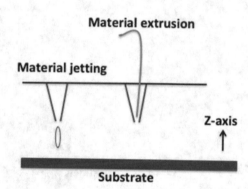

surfaces of the substrate. These nozzles represent either different processes or the same process. Nozzles are for extrusion or jetting or cold spraying.

Even if three nozzles represent three different processes, these processes are not combined or related to each other. These nozzles act on different surfaces of the substrate. Depositing by one nozzle is not affected by the working of other nozzles. Thus these processes, without being compatible with each other, without synchronizing the movement of one nozzle with respect to another nozzle, can be used in the system.

Considering two of these nozzles are equipped in the system to build in z-direction (Fig. 7.16). In other words, two nozzles are used to modify the same surface or same plane of the substrate. If these two nozzles represent two different processes, for example, one is for material jetting and other is for material extrusion. The results of two processes need to be compatible with each other to make a defect-free deposition. For example, if half plane or half layer is made by jetting and other half layer is made by the extrusion, the deposition that comes from either nozzle should give rise to the same height of the deposition. If the height is not same after the solidification of the materials, the results of these two processes are not compatible with each other.

If materials deposited from either nozzle do not adhere with each other giving rise to weak bond between two depositions which can be leading to separation of two depositions from each other with ease, or the boundary between two depositions is weak enough to let the depositions detached from each other, two processes are not compatible with each other.

To make these processes compatible, right materials need to be chosen for both processes. It means whatever materials available for either process are not readily available to be used, these materials need to go through testing so that the right materials can be selected from the available materials. Thus the materials available to be used by either process become limited. This limitation of materials is restricting either process not to work with their best materials if their best materials are not among those few selected materials. This limitation of materials is restricting the performance of both processes. The processes are expected to give up their best performance only because the processes need to be compatible. One process is interfering in the business of the other process only because they have to become compatible.

Combining two processes demands compatibility. What if processes refuses to be compatible— it brings a question why these processes are brought together to be compatible. Both processes work with different sets of materials, while material in material jetting should have low viscosity the material in material extrusion should have high viscosity so that extruded material will not be disintegrated into small drops similar to what is achieved with material jetting.

While material jetting works with liquid feedstock, material extrusion works with solid feedstock. While material extrusion requires a heating device for the conversion of its solid feedstock into extrudable material, the material jetting does not require such heating device. While extruded material when deposited almost refuses to lose its shape and the extruded shape is almost visible, the jetted drop almost refuses to have the shape of the drop when it touches the substrate. While the surface tension of the substrate governs the shape of jetted material, the same surface tension of the substrate becomes helpless when the extruded material comes. While jetted material shrinks high because it needs to solidify from liquid state, the extruded material shrinks low because it solidifies only from semi-solid state.

There can be some materials which will not fit in the above generalization, but these few materials will not help have an adequate pool of materials to choose from. For example, some type of jetting printhead can deposit droplets using high-viscous material. In this case, both processes can be used for the material of the high viscosity.

This type of printhead is expensive and its use is not a common practice. If it was common practice, material jetting system would be too expensive to survive in the face of inexpensive material extrusion system for making low-value products.

If the material needs to be developed for material jetting, material goes through one transition from liquid to solid state starting from feedstock stage to the fabrication stage. While in material extrusion, the material goes through two transitions: from solid to semi-liquid and then to solid state.

3.7 Process and Processness

Instead of finding material properties compatible so that processes can become compatible, it is possible that processes can itself be modified so that a process loses its processness. In material extrusion, if the heater is used to heat more, the material starts to lose its viscosity, there comes a point, when the extruded material starts to lose its extrudability and starts to gain jettability. So that the extruded material is no longer a long semi-solid thick line waiting to be laid on a substrate but the extruded material is a streamline liquid flow waiting to be fragmented into several drops if there is a slight variation in the process condition. Similarly, in material jetting, if a liquid of higher viscosity is chosen, it does not come out in the form of drops but a long extended drop resembling to an extruded material that has lost almost all of its extrudability.

Thus in both cases, processes are modified so that both processes march towards each other continuously losing their identities in the course of movement. If a process loses its identity, there is no rationale left anymore for selecting the process. Two processes are brought together because they are different. They are different because they work with different sets of principles, having a material which cannot be processed by one but can be processed by other, having an application which can be achieved with one process but cannot be achieved with other. When they are brought together, it is expected both their advantages will meet to give a unique product that is not achievable by either process (which is different). If both processes become compatible by modifying them, there is no original process left anymore.

3.8 Using Same Process

One process can sometimes help achieve the properties that can be achieved by using two processes combined. If an additional process is used so that it deposits a material which gives different properties, there is no need of two different processes, one process can do the job by using two outlets. Figure 7.17 shows two nozzles using the same process are fitted with different materials to get different properties at different sections of the same layer [27]. If two different structures at different section of the same layer are required, the same process can still do it by changing parameters. It does not mean a combination of processes has no merit, it only means one process can help achieve more properties if the potential of the process is realized (refer Chap. 9).

Fig. 7.17 One process
using two nozzles to
deposit different materials

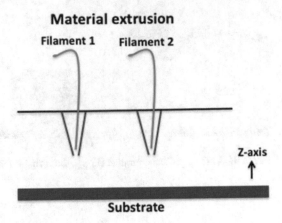

3.9   Many Systems Instead of One System

Figure 7.15b shows three nozzles that have deposited on three different surfaces of a substrate. These three nozzles represent three different processes or just one process but the presence of three nozzles in a system does not mean the presence of three nozzles is required to be in the same system for the substrate to be modified (Fig. 7.15b) and the product to be formed (Fig. 7.15c).

If three nozzles are fitted in three different systems (system 1, 2, 3) without changing their respective positions in the systems, again the same product (Fig. 7.15c) can be fabricated by serially working on the first surface by the first nozzle in the first system and then on the second surface by the second nozzle in the second system and so on, by taking the substrate from one system to the next system (Fig. 7.18).

Thus the presence of all nozzles in one system instead of three separate systems means the advancement of a system so that three nozzles can be fitted but is not the advancement of compatibility of processes. Since there is more problem in the development of combined process than the development of combined system, it will be the right strategy to develop advanced systems without developing the combined process.

If there is a single nozzle as in Fig. 7.1a, only the horizontal surface of the pre-product or the substrate can be modified. In this case, even in the presence of well-compatible pre-product, there will be limitation in the complex product fabrication. Even by changing the orientation of the pre-product, there will still be limitation. Changing the orientation is not possible in all types of pre-products. Moreover, if a pre-product is fabricated with an aim to convert it in a complex product by changing its orientation, this aim will itself limit the complexity of the pre-product that can be fabricated.

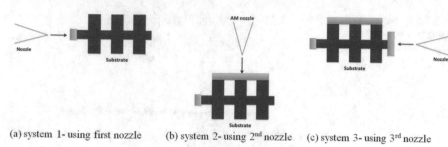

(a) system 1- using first nozzle   (b) system 2- using 2nd nozzle   (c) system 3- using 3rd nozzle

Fig. 7.18 Making the same product (Fig. 7.15c) using three systems in three successive steps

3.10 Inability of AM Systems

Majority of commercial systems are having a single nozzle or many nozzles to work only on horizontal surfaces (a single build direction) implying the horizontal surface is a preferred surface to build an object on it. Irrespective of which process these nozzles are representing, the surfaces being horizontal remain unchanged. It does not mean a process is never tried to work on other surfaces, they are tried in repair or in a recess zone or in some lab settings, but this is not the norm. If a process is developed, it means the process is developed with respect to only a horizontal surface, a developed process has not become a developed process because it has proved its capability to work on other than horizontal surfaces. Not a single commercial AM system is available which will be able to make a single product if the system is turned vertically by 90°.

3.11 Plastic System

Horizontal surface being a preferred surface is not without reason. On non-horizontal surfaces, there requires a mechanism to counteract gravity, otherwise drops will roll down or extruded track shapes will distort. If the non-horizontal surface is a vertical surface as shown in Figs. 7.15, 7.18, drops do not have even chance to roll down. They fall if they do not stick to the surface, thus extruded track is unlikely to stick.

In material extrusion, it is the filament the forward movement of which applies pressure on the semi-molten material to be extruded. The role of pressure by the filament is to maintain continuous onward supply of the extruded material. If the deposition needs to occur on a vertical surface, the filament needs to apply more pressure lest the extruded material sags and not be able to reach the surface. The filament cannot apply more pressure, if the filament is expected to apply more pressure, filament should be stiff. If the filament becomes stiff, it does not have enough pliability to travel from the supply to the nozzle.

But, filament is not an essential requirement of material extrusion process [28, 29]. If filament cannot do the job, others can do the job, the process is still a material extrusion process as long as there is extrusion taking place. A filament plays dual role—for applying pressure and a source of materials. If there is a material extrusion system where a piston replaces the filament for applying pressure and a reservoir of materials replaces the filament as a source of materials, this non-filament based extrusion system can be tried to deposit on a vertical surface.

The nozzle needs to be near to the vertical surface so that when the piston applies pressure the extruded material does not need to travel far. High pressure is required because the extruded material remain in a high-viscous state. High pressure will project the material on the surface otherwise the material will not reach and adhere.

The projected material is no longer the extruded material. It is inherent in the term extrusion that the extruded material (extrudate) is that material which takes shape due to the gravity as a dominant cause, which is only possible if the material falls along the gravity. If, for example, the system is configured in an inverted position so that the nozzle extrudes materials upward on a substrate kept above it, the extrudate will not reach to the substrate and fall on the nozzle itself. If the material is extruded with high force so that it can reach the substrate, the material is no longer an extruded material, the material has become a projected material. The shape and trajectory of the projected material are no longer the shape and trajectory of the extruded material because the projected material is due to high ejection force plus gravity while the extruded material is due to low ejection force plus gravity.

Thus using material extrusion system and depositing on a vertical surface does not mean material extrusion is taking place. It does not mean whatever deposition is taking place should be stopped and discarded because the deposition is no longer fulfilling the criteria of a particular process. It only means the processes developed and used on horizontal surface will have no guarantee to work on a vertical surface, new processes need to be developed for vertical surface.

Material jetting goes through jetting or projecting liquid drops and therefore is better placed than material extrusion to deposit on a vertical surface. Thus in this process, there is guarantee that the drop will reach to the vertical surface without the need to apply additional pressure on liquid. But, the pressure applied on the liquid has limited objective, the pressure is applied to guide the drops to the right place so the drops will not change its location inadvertently, the pressure is not applied to cause adherence to the horizontal surface or to the deposited tracks on the surface.

If the process is used to deposit on a vertical surface, drops are jetted or projected with high pressure. Such jetting is used if forcing the drop to deform and get attached by anchoring when the drop strikes the surface is a mechanism. But this will not work because such low-viscous drops will splash and be fragmented when they strike the surface with force. High-viscous liquid will not be having such problems. High viscosity is not enough, the drop must get attached to the surface, the material must have stickiness for a particular surface. This means the material worked for horizontal surface will not have usability for the vertical surface. New material needs to be developed, the material will not work for the same condition of the surface, the surface having high roughness will provide more anchor. This means that any type of pre-product will not be fitted even if there is a new material to work on. This brings limitation of the pre-product that can be used.

Thus development of a new material, tailoring the pre-product that will suit the new material, bringing the new material to the various types of non-horizontal surfaces, guiding these materials to stick not only on the surface but also on previously attached layer, making an extended structure defying gravity, will lead to the development of a process.

But this development of a process is not the development of a combination of processes (that development). This development does not demand compatibility between two processes which is demanded in that development. This development is an ongoing development while that development is a radical departure. This development is a less team work while that development requires a team as a necessity. This development can be uncertain because the development is not sufficient while that development will be uncertain because that development depends on this development.

Thus the development of a process plus manufacturing strategy is free from many problems that come with the development of a combination of processes plus manufacturing strategy.

3.12   Metal System

While for making plastic products, there are no AM processes available which will work on vertical surfaces. For making metallic products, some processes (cold spray AM, laser powder DP) are available.

Cold spray AM (CSAM) deposits by projecting high velocity particles on the surface [30]. This high velocity distinguishes CSAM from other powder DP [31]. The high velocity gives the particle necessary kinetic energy to be converted later into potential energy on striking the surface, if surface material is ductile, the particle does not ricochet but deforms and attaches to the surface. If the surface material is not ductile, it requires to be heated to become ductile. Thus, a particle in CSAM is getting attached without being melted. If it is not melted, its original texture does not change. This is the reason why CSAM is better than other DP.

But the very reason for which it is better is also the reason why it is not able to make fine features. High velocity impact will break any fine feature that is going to be built, thus the process is in direct conflict with fine features. Avoiding high velocity impact is not a solution but will create another problem because it is the high impact which refines the grain and increases the strength [32]. CSAM also restricts the properties (ductility) a target or surface material can have, therefore all types of pre-products will not be suitable if this process is supposed to modify the vertical surface.

Powder DP relies on melting of powders for getting bonded to the vertical surface. In this process, powders reach the surface without impact and get melted while in contact with the surface. The molten drop or melted powder does not slide from the surface because it is anchored on the surface due to surface tension of the surface. If the surface tension is low, the melted powder will fall or if the weight of melted powder is high, surface tension will not able to hold it. The powder gets melted when it comes in contact with the hot surface. If the surface is not hot, the powder will not be melted while if the surface is hot, its surface tension changes. Thus a balance is required depending upon materials, between its two duties—a duty to melt and a duty to hold.

In this process, there is no such impact by powder on the surface, thus this process is not prone to break any fine features by impact as CSAM is. Thus the fabrication of fine features will not be prevented by this process for the same reason it is prevented in CSAM. But this process is not free from problems: there are molten materials, there is sliding, and there is a need to hold back the deposited molten material accurately which will bring limitation to the fineness of a feature.

3.13   Pre-Product Compatibility

Making fine features is not the realm of DP. Figure 7.15a shows a system having provision for only DP, thus this system is not expected to add complexity on a pre-product.

If a system is equipped with bed process (BP) as well, the pre-product is not deprived from the advantages given by BP. But for a pre-product to be able to be fitted in BS, the pre-product itself should be deprived from the complexity. A pre-product which is complex enough that its top surface consists of features of unequal heights will not allow a bed of uniform thickness to be formed on its top surface (Fig. 7.19a). Similarly, a pre-product which is complex enough that its bottom surface consists of features of unequal heights will not allow a bed of uniform thickness to be formed on its top surface (Fig. 7.19b), the pre-product will be tilted because of unequal heights.

Thus complexity of a pre-product with respect to BS means a part having features of unequal heights. It is possible that a part is simple rectangular but is complex because of two walls (1, 2 in Fig. 7.20a) of unequal height on it while a complex part having a number of complex features (3, 4, 5 of Fig. 7.20b) of equal height is non-complex. Further, a part having all features (6, 7 of Fig. 7.20c) having equal height is again non-complex.

(a) pre-product having unequal features on top surfaces

(b) pre-product having unequal features bottom surfaces

Fig. 7.19 Complex pre-products due to features of unequal height on its top (**a**) and bottom (**b**) surfaces

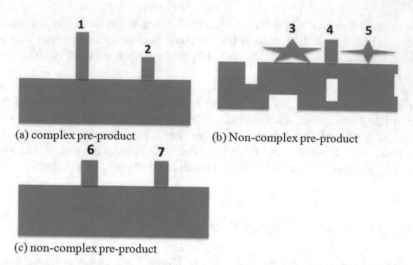

(a) complex pre-product   (b) Non-complex pre-product

(c) non-complex pre-product

Fig. 7.20 Complex (**a**) and non-complex (**b**) pre-product with respect to their fittability in BS

A pre-product that can be fitted in BS is non-complex while that cannot be fitted is complex. From fittability point of view, a part is complex that is not complex when it was made (Fig. 7.20a). On the contrary, a part is not complex that is complex when it was made (Fig. 7.20b).

This brings a question why a part or pre-product needs to fitted in BS—the answer is simple a part needs to be fitted because it is assumed that when the part will be fitted, BP will make a complex feature on it that no other process can.

A simple part needs to be taken to perform the reality check. Figure 7.20c shows a part which is two times simple—once when it was made and the second time when it can be fitted. This part when fitted in BS allows complex features to be made only on two of its features, i.e., 6 and 7. Thus, whatever complexity is added on the part should be added on only two features.

But what about the vast area of the part adjacent to or between feature 6 and 7, these areas are out of reach of the system. Those features, which reach to the level of bed, are the only features which can be or cannot be built upon it. The action taken by BP becomes null and void beneath the boundary of bed—if a beam is taking action, the beam will not cross the boundary and if beam crosses the boundary, consolidation will not take place and if some consolidation takes place, it cannot be controllable.

If BS is not allowing to build on the vast area, it does not mean the system will also allow to bypass its rule and allow to build over the vast area by another method. Another method means building a section on feature 6 or 7 and extending the section towards the vast area so that if a section is not built on the vast area, at least some section will be made overlooking the vast area. But BS does not even allow to make any section overlooking it. It is not only the vast area that is a forbidden zone but the free space over the vast area is also a forbidden zone.

What if the attempt is done to build over the vast area connecting to the feature 6 or 7, the building relies on the material (powder, photopolymer, slurry) to provide support to ongoing built. The built is an overhang feature, it has thus all the shortcomings of an overhang. The built does not have the advantage to be supported by support structure (SS) because it cannot be made. Arguing that the fabrication of such inferior overhang is better than no overhang features at all, therefore there is no harm to be benefitted by this inferior advantage offered by BS. But the acceptance of this offer relies on the assumption that BS is always better for making complex parts ignoring the fact that BS is not always better when a part is not suitable to be fitted into it.

3.14 Overhang in Combined System

If the same part is tried in deposition system (DS), the vast area can be accessed and an overhang can be made because SS can be made as well to support the overhang. What cannot be done in BS can be accomplished in DS. But SS can be made anywhere, if the area between the feature 6 and 7 is narrow, SS cannot be made between 6 and 7. Therefore, an overhang cannot be made. If DS is not able to make SS, the overhang required at that place is not made by AM anymore. It brings limitation to the pre-product that can be fabricated. If fabricated, that can be fitted; if fabricated and fitted, that can be converted into a product.

Thus DS is better than BS when SS is required to make an overhang on a pre-product. When SS is made and overhang is made on it, what will happen next, what the DS can do further. DS has limitation to make a complex overhang or to add enough complexity on the overhang. After the fabrication of overhang, DS cannot go far. If BS was able to make overhangs, BS could add more complexity on it than DS.

DS made the overhang, that is why DS is better than BS, but this was due to situational compulsion of BS when a non-suitable pre-product was fitted in BS. Expecting BS to adjust with a pre-product exposed the inherent limitation of BS. But DS has not such limitation because DS is rather an open system. After the trial of overhang is over, DS needs to be compared with BS again. Fabrication of overhang is the fabrication of base. If the base is available, BS is always better than DS because BS can add more complexity on the base or overhang.

3.15 Combination of Both Systems

Why not both systems (DS and BS) should be combined—one system will make overhang while other will work on that overhang, one system provides base while other utilizes that base.

It is the combination of two systems rather than combination of two processes. When one system finishes the work, other system starts to work. No compatibility between both processes is required. Combining two systems is not new, they are combined to create weak SS in metal PBF [21].

Figure 7.21 shows a combination of three DS and one BS. A pre-product which cannot be fabricated in BS is shown in Fig. 7.22a. Overhang needs to made on the

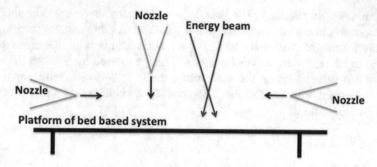

Fig. 7.21- Combination of BS and DS

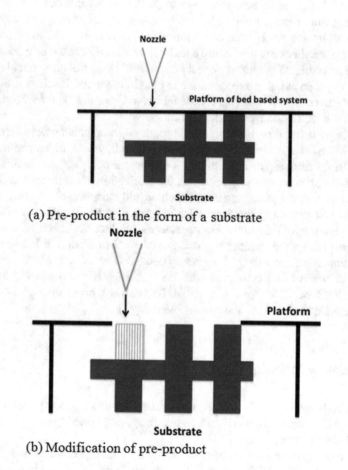

(a) Pre-product in the form of a substrate

(b) Modification of pre-product

Fig. 7.22 Modification of pre-product by DS

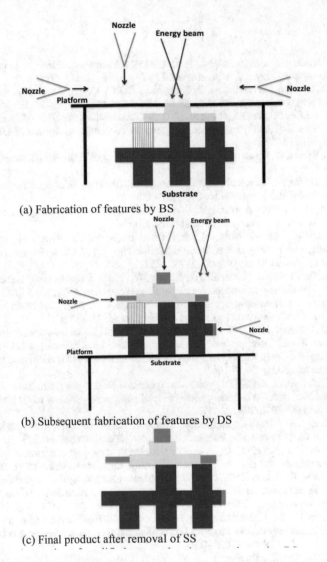

(a) Fabrication of features by BS

(b) Subsequent fabrication of features by DS

(c) Final product after removal of SS

Fig. 7.23 Conversion of modified pre-product into a product using BS and DS

pre-product but due to lack of support SS, fabrication of overhang is not possible. SS was made by a nozzle while pre-product was fitted in BS (Fig. 7.22b). After the fabrication of SS, BS was used to utilize SS to make an overhang on it (Fig. 7.23a). Further modification of the pre-product was done by bringing it above the platform (Fig. 7.23b). After modification, SS was removed (Fig. 7.23c).

Thus a product (Fig. 7.23c) is made using a fabrication strategy combining three processes: two AM processes and another process (e.g., machining) for making the pre-product. There could be more processes if the required product was more complex. There could be pre-process and post-process as well if it was necessary for the product development [33].

References

1. Ituarte, I. F., Coatanea, E., Salmi, M., et al. (2015). Additive manufacturing in production: A study case applying technical requirements. *Phy Procedia, 78*, 357–366.
2. Diegel, O., Schutte, J., Ferreira, A., & Chan, Y. L. (2020). Design for additive manufacturing process for a lightweight hydraulic manifold. *Additive Manufacturing, 36*, 101446.
3. Didier, P., Le Coz, G., Robin, G., et al. (2021). Consideration of SLM additive manufacturing supports on the stability of flexible structures in finish milling. *Journal of Manufacturing Processes, 62*, 213–220.
4. Martin, J., Yahata, B., Hundley, J., et al. (2017). 3D printing of high-strength aluminium alloys. *Nature, 549*, 365–369.
5. Nelson, J., Berlin, A., Menold, J., & Parkinson, M. (2020). The role of digital prototyping tools in learning factories. *Procedia Manuf, 45*, 528–533.
6. Diegel, O., et al. (2019). *A practical guide to Design for Additive Manufacturing. Springer series in advanced manufacturing*. Singapore: Springer.
7. Friesike, S., Flath, C. M., Wirth, M., & Thiesse, F. (2019). Creativity and productivity in product design for additive manufacturing: Mechanisms and platform outcomes of remixing. *Journal of Operations Management, 65*, 735–752.
8. Gäumann, M., Bezençon, C., Canalis, P., & Kurz, W. (2001). Single-crystal laser deposition of superalloys: Processing–microstructure maps. *Acta Materialia, 49*(6), 1051–1062.
9. Sreeramagiri, P., Bhagavatam, A., Alrehaili, H., & Dinda, G. (2020). Direct laser metal deposition of René 108 single crystal superalloy. *J Alloy Compd, 838*, 155634.
10. Blok, L. G., Longana, M. L., Yu, H., & Woods, B. K. S. (2018). An investigation into 3D printing of fibre reinforced thermoplastic composites. *Additive Manufacturing, 22*, 176–186.
11. Xu, Y., Wang, Z., Gong, S., & Chen, Y. (2021). Reusable support for additive manufacturing. *Additive Manufacturing, 39*, 101840.
12. Chou, YS., & Cooper, K. (2017). Systems and methods for designing and fabricating contact-free support structures for overhang geometries of parts in powder-bed metal additive manufacturing. US patent 9767224.
13. Calignano, F. (2014). Design optimization of supports for overhanging structures in aluminum and titanium alloys by selective laser melting. *Materials and Design, 64*, 203–213.
14. Cheng, B., & Chou, K. (2015). Geometric consideration of support structures in part overhang fabrications by electron beam additive manufacturing. *Computer-Aided Design, 69*, 102–111.
15. Zhao, G., Zhou, C., & Das, S. (2015). Solid mechanics based design and optimization for support structure generation in Stereolithography based additive manufacturing. In *Volume 1A: 35th Computers and Information in Engineering Conference*.
16. Lušić, M., Feuerstein, F., Morina, D., & Hornfeck, R. (2016). Fluid-based removal of inner support structures manufactured by fused deposition modeling: An investigation on factors of influence. *Procedia CIRP, 41*, 1033–1038.
17. Bobbio, L. D., Qin, S., Dunbar, A., et al. (2017). Characterization of the strength of support structures used in powder bed fusion additive manufacturing of Ti-6Al-4V. *Additive Manufacturing, 14*, 60–68.
18. Vaissier, B., Pernot, J. P., Chougrani, L., & Véron, P. (2019). Genetic-algorithm based framework for lattice support structure optimization in additive manufacturing. *Computer-Aided Design, 110*, 11–23.
19. McConaha, M., Venugopal, V., & Anand, S. (2020). Integration of machine tool accessibility of support structures with topology optimization for additive manufacturing. *Procedia Manuf, 48*, 634–642.
20. Ni, F., Wang, G., & Zhao, H. (2017). Fabrication of water-soluble poly(vinyl alcohol)-based composites with improved thermal behavior for potential three-dimensional printing application. *J Appl Poly Sci, 134*, 44966.

21. Wei, C., Chueh, Y., Zhang, X., et al. (2019). Easy-to-remove composite support material and procedure in additive manufacturing of metallic components using multiple material laser-based powder bed fusion. *ASME J Manuf Sci Eng, 141*(7), 071002.
22. Chan, Y. L., Diegel, O., & Xu, X. (2021). A machined substrate hybrid additive manufacturing strategy for injection moulding inserts. *International Journal of Advanced Manufacturing Technology, 112*, 577–588.
23. Bambach, M., Sizova, I., Sydow, B., et al. (2020). Hybrid manufacturing of components from Ti-6Al-4V by metal forming and wire-arc additive manufacturing. *Journal of Materials Processing Technology, 282*, 116689.
24. Silva, M., Felismina, R., Mateus, A., et al. (2017). Application of a hybrid additive manufacturing methodology to produce a metal/polymer customized dental implant. *Procedia Manuf, 12*, 150–155.
25. Godec, M., Malej, S., Feizpour, D., et al. (2021). Hybrid additive manufacturing of Inconel 718 for future space applications. *Materials Characterization, 172*, 110842.
26. Merklein, M., Junker, D., Schaub, A., & Neubauer, F. (2016). Hybrid additive manufacturing technologies – An analysis regarding potentials and applications. *Phy Procedia, 83*, 549–559.
27. Watschke, H., Waalkes, L., Schumacher, C., & Vietor, T. (2018). Development of novel test specimens for characterization of multi-material parts manufactured by material extrusion. *Applied Sciences, 8*(8), 1220.
28. Volpato, N., Kretschek, D., Foggiatto, J., & Cruz, C. (2015). Experimental analysis of an extrusion system for additive manufacturing based on polymer pellets. *International Journal of Advanced Manufacturing Technology, 81*(9), 1–13.
29. Kishore, V., Ajinjeru, C., Nycz, A., et al. (2017). Infrared preheating to improve interlayer strength of big area additive manufacturing (BAAM) components. *Additive Manufacturing, 14*, 7–12.
30. Yin, S., Cavaliere, P., Aldwell, B., et al. (2018). Cold spray additive manufacturing and repair: Fundamentals and applications. *Additive Manufacturing, 21*, 628–650.
31. Paul, C. P., Mishra, S. K., Kumar, A., & Kukreja, L. M. (2013). Laser rapid manufacturing on vertical surfaces: Analytical and experimental studies. *Surface and Coating Technology, 224*, 18–28.
32. Xie, X., Ma, Y., Chen, C., et al. (2020). Cold spray additive manufacturing of metal matrix composites (MMCs) using a novel nano-TiB2-reinforced 7075Al powder. *J Alloy Compound, 819*, 152962.
33. Biondani, F. G., Bissacco, G., Mohanty, S., et al. (2020). Multi-metal additive manufacturing process chain for optical quality mold generation. *Journal of Materials Processing Technology, 277*, 116451.

Chapter 8
Mass Production

1 Is Mass Production Possible Using an AM System?

AM allows mostly few parts to be produced but all types of production such as single part or job production, few hundred parts or single batch production [1], around thousand parts or mass production are possible. In comparison to million parts produced in conventional manufacturing (CM), few thousand parts is still a small number to be called a mass production. But since this number is more than few hundred parts which is single batch production, this number can be a lower end of mass production.

Not all AM systems allow even that lower end of mass production to take place. If a part is small while the system is big, thousand parts can be fabricated (Fig. 8.1c). Such system can be a bed system (BS) (such as powder or photopolymer bed), which allows a part to be fabricated over other parts albeit with some gap between them [2]. Though questions arise whether mass production should be the aim or can be cost-effective or energy-efficient or material-efficient in comparison to CM, but it can be technically feasible using one AM system. Combining this number advantage with design advantage of AM, mass production can be taken to another level, i.e., mass customization [3].

1.1 Methods for Production

Commercial AM systems produce parts serially, i.e., through two types: (1) the system starts making a second part after it finishes the fabrication of the first part (Fig. 8.1a) and (2) the system starts scanning the first layer of a second part after it finishes scanning the first layer of the first part (Figs. 8.1b, 8.2a). Figure 8.1a shows only one part in the bed, the second and successive parts will be formed one part at

© The Author(s), under exclusive license to Springer Nature Switzerland AG 2022
S. Kumar, *Additive Manufacturing Solutions*,
https://doi.org/10.1007/978-3-030-80783-2_8

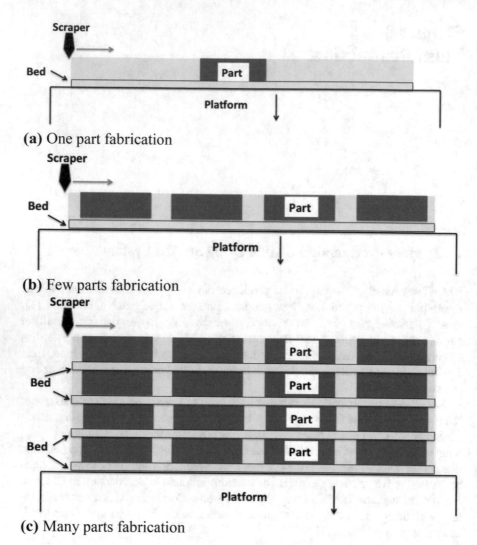

Fig. 8.1 Methods to produce a number of parts in bed process (BP): (**a**) one part fabrication, (**b**) few parts fabrication, (**c**) many up to thousand parts fabrication

a time in the bed. Figure 8.1b shows four parts in the bed, the next four parts will be fabricated after the fabrication of these four parts and so on. The first type is akin to job production while the second type is akin to small batch production. The second type allows many parts to be fabricated without creating a new bed over a part. Figure 8.1c shows the creation of a bed over a batch of parts for three batches. Figure 8.1c demonstrates that 16 parts can be fabricated in an AM system, the next 16 parts can be fabricated again in the same way, and so on.

1.2   Serial and Non-serial Production

The second type though does not fabricate a part after finishing another part, but it still comes under serial production because it finishes some portion of a part and then starts fabricating some portion of another part and so on (Fig. 8.2)—thus processes parts serially. It will not be a serial production if nth layer (any layer such as first, second, third, etc.) of all parts are processed at the same time (Fig. 8.3). Non-serial production is possible if processing does not move from nth layer of one part to nth layer of another but nth layer of all parts will be scanned simultaneously by arranging the number of beams equal to the number of parts (one beam for each nth layer) (Fig. 8.3). Again, non-serial production is also possible by exposing or irradiating nth layer of all parts instantly either by a number of beams or sources (Fig. 8.3) or by a bigger beam or source (Fig. 8.3).

In serial production, if scanning a layer takes 5 min, scanning a layer of four parts will take $5 \times 4 = 20$ min (Fig. 8.2a). If instead of raster scanning a layer, irradiating a layer takes 1 s, irradiating a layer of all four parts will take $1 \times 4 = 4$ s (Fig. 8.2b). In non-serial production, if scanning a layer takes 5 min, scanning a

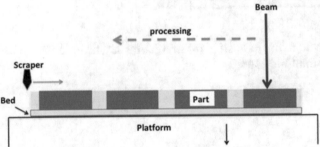

(a) Serial production by scanning nth layer of a part after nth layer of another

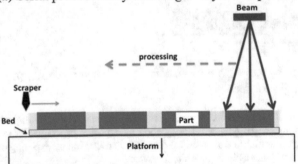

(b) Serial production by instantly irradiating nth layer of a part after nth layer of another

Fig. 8.2 Serial production in BP. (**a**) Serial production by scanning nth layer of a part after nth layer of another. (**b**) Serial production by instantly irradiating nth layer of a part after nth layer of another

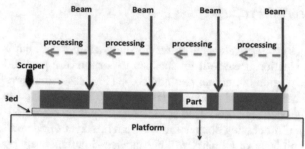

(a) Non-serial production by scanning nth layer of all parts simultaneously using multiple sources

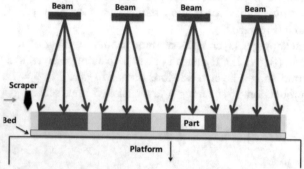

(b) Non-serial production by exposing or irradiating nth layer of all parts simultaneously using multiple sources

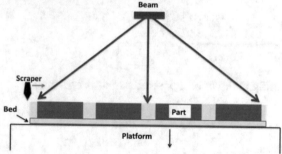

(c) Non-serial production by exposing or irradiating nth layer of all parts simultaneously using a single source

Fig. 8.3 Non-serial production in BP. (**a**) Non-serial production by scanning nth layer of all parts simultaneously using multiple sources. (**b**) Non-serial production by exposing or irradiating nth layer of all parts simultaneously using multiple sources. (**c**) Non-serial production by exposing or irradiating nth layer of all parts simultaneously using a single source

layer of all four parts takes only 5 min (Fig. 8.3a). Again, in non-serial production where processing takes place by irradiating all nth layers at the same instant (Fig. 8.3b, c), processing nth layer of all parts will take the time equal to the time of irradiation, it can be millisecond or second.

If processing moves from one part to another in a bed to process the same ordinal number of layers, the same step (processing nth layer) of production is done one by one—it typically comes under small batch production. For the fabrication to come under mass production, the same ordinal number of layers (for example, nth layer) need to be processed at the same time so that the same step of production is executed at the same time (Fig. 8.3). Thus, if thousand parts are produced in present-day commercial AM systems (which are serial production systems), it can be called mass production implying the production is in mass scale but it cannot be mass production technically because this mass production is technically different from the mass production in CM.

1.3   Fast Serial Production

Arcam multi-beam processing is an attempt in this direction where several beams process the bed at the same time but this attempt is more to improve the fabrication efficiency of a part than to process multiple parts. There is no commercial AM system yet available which fabricates a part by exposing whole bed by en energy source; therefore, there is no commercial AM system which fabricates parts in mass production fashion. Area scanning by digital micromirror device (DMD) fabricates a part by exposing some area of the bed, it is somewhat similar to the concept of parallel scanning of multiple parts in a bed if the exposing covers bigger area of the bed. DMD is mostly used for fabricating microparts [4, 5].

In laser powder bed fusion (LPBF), using two beams to scan adjacent parts leads to the deterioration of the quality of parts due to the plume [6] or melt pool [7] generated by a beam. Thus two parts cannot be expected to be fabricated very close to each other. Making a large number of parts means dealing with an increased amount of vapor and vapor-induced defects in parts [8].

Thus area scanning by DMD fabricates parts by exposing some portion of the bed, then exposing another portion of the bed. This type of scanning or exposing continues until whole bed is covered. If there is a big part (to be fabricated) which covers all area of the bed, it is fabricated by multiple-exposing of the same part by moving from one portion to another. If there are many small parts (to be fabricated) which cover all area of the bed, they are fabricated by multiple-exposing of the same bed by moving from one part to another part (similar to shown in Fig. 8.2b). Since there is a movement of processing from one part to another in the same bed, it is again a serial production.

1.4   Non-serial Production

Imagining a state where area exposed by DMD is large enough to cover whole bed so that there is no need for any movement of processing from one area to another area of the bed (Fig. 8.3c). Then whatever number of parts will be arranged to be formed in a bed, all will be irradiated at the same time without any movement of the processing. This is no longer a serial production but a non-serial production which is similar to mass production.

Thus serial and non-serial production differs in how bed is exposed—if the whole bed is exposed simultaneously so that all parts grow simultaneously or if the whole bed is exposed partially plus partially so that all parts do not grow simultaneously. Thus the difference between serial and non-serial production lies in what happens for one set of parts to be built in a bed. After one set of parts is already built, another set of parts needs to be built by creating an intermediate bed by moving in the build direction (Fig. 8.1c)—this movement from one set to another set is same in both serial and non-serial production. This movement is a serial movement. This serial movement is a necessity if production in mass scale is to be realized. Non-serial production can technically be a mass production but cannot achieve mass production level if production volume is not accumulated serially by producing serially with serial movement in the build direction.

This brings a question why not non-serial production should be considered serial production because it is the serial component of the non-serial production process that gives bigger contribution leading to mass production. It is because mass production in all types of manufacturing is a result of serialwise production. For example, mass production in injection molding (IM) is realized by producing serially in a batch consisting of few parts. But IM is mass production because these few parts in a batch are produced simultaneously and not one by one.

If all parts are produced by instant simultaneous irradiation (Fig. 8.3b, c), the production will be faster than if all parts are produced by serial irradiation (Fig. 8.2b). Thus, non-serial production will be faster than the serial production. But, what if simultaneous irradiating or scanning all parts (Fig. 8.3) does not turn out to be faster than the serial irradiating or scanning (Fig. 8.2)—even then non-serial production rather than serial production is technically similar to mass production.

1.5   Number of Parts

Considering BS having a working dimension of 1 m × 1 m × 1 m. If the size of a part is 1 cm × 1 cm × 1 cm, at least $60 \times 60 = 3600$ parts can be accommodated in x-y plane in a bed (Fig. 8.4). In this example, the size of the part is kept small while the size of the system is kept large. For mass production to occur, there should be considerable difference between the sizes of the part and the system. Therefore, bigger part in a small system is already excluded for realizing the mass production. The size of the system cannot be arbitrarily increased because it will be difficult to maintain uniform condition in a big system. Therefore, it will be better to have two systems rather than one big system if it is difficult to maintain uniform condition due to the big size of the system. Hence, in this example, the size of the system is not kept larger than 1 m × 1 m × 1 m.

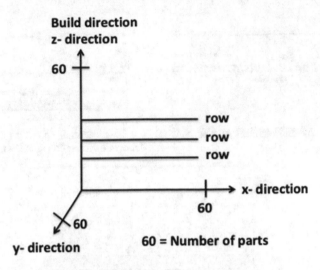

Fig. 8.4 Arrangement of parts to be fabricated in BS

In this example, more than 3600 parts can be accommodated if adequate space is not given between any two parts to minimize the effect of the fabrication of one part over adjacent parts. These 3600 parts will occupy a height of slightly more than 1 cm covering whole working space on the bed. There is still a space up to a height (in z-direction or build direction) of 1 m available for parts to occupy if they are arranged vertically in several rows. There can be at least 60 rows of parts, the number is not kept high so that the quality of parts in one row is not affected during the fabrication of the next row. Thus, a total of 3600 × 60, i.e., 216,000 parts can be fabricated in an AM system in one go (Fig. 8.4). If the size of the system is smaller and the size of the part is bigger, still few thousand parts can be fabricated.

Few thousand parts are not a large number that they can be compared with higher end of mass production of CM but they are still in sufficient number and can be compared with lower end of mass production of CM. If this number of parts need to be made, AM can still have advantage over CM.

1.6   Processing Space and Machine Space

In BS, processing space and machine space are different (Fig. 8.5). Processing space means the volume where processing happens or where parts are fabricated from material or where material interacts with an energy beam. This space is the space of a processing chamber. Processing space can also be called as working space. Figure 8.5a shows the position of the platform before the start of the fabrication in BS. Bed is formed where parts will be formed. Since no part or layer of the part is fabricated, the position of the platform does not go down to give space for the

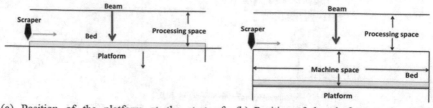

(a) Position of the platform at the start of fabrication

(b) Position of the platform at the end of fabrication (parts are not shown)

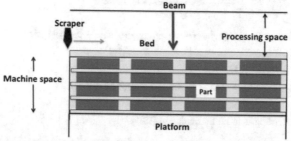

(c) Position of the platform at the end of fabrication (parts are shown)

Fig. 8.5 Processing space and machine space in BS. (**a**) Position of the platform at the start of fabrication. (**b**) Position of the platform at the end of fabrication (parts are not shown). (**c**) Position of the platform at the end of fabrication (parts are shown)

part. Figure 8.5b, c show that the platform is gone down giving rise to a space, this space is called machine space or build volume. Though parts are built in z-direction, the parts occupy the space in −z-direction.

Thus machine space is the place where parts are gradually getting stored during fabrication. When parts are getting made layer by layer, the parts are getting slipped layer by layer in the machine space. Machine space is also known as build volume, but the build volume actually implies a storing volume rather than a space where fabrication happens. The name machine space is more appropriate because this space or storing volume determines the capacity of the machine. This space determines the maximum size of the part that can be made, or maximum number of parts that can be fabricated. This is the space because of which one commercial machine becomes smaller or bigger. Thus the maximum size of the part or maximum number of parts does not depend upon the processing space.

The processing space is more to ensure that processing happens without any problem. Thus the processing space requires only that minimum space which will ensure the happening of the processing. The processing space does not require more space because it does not let the processed part either fully or partially remains in its space. Thus the building of a part does not tend to occupy the processing space. Again, in the course of occupying, it does not try to shrink the further processing space. Thus the processing space is free from the responsibility to hoard the parts after the fabrication. The processing space has limited duty, this limitation in duty enables the processing space always to remain constant in dimension. This implies that by keeping the processing space constant, the number of parts can be increased.

What processing space wants. The space only wants that the vertical distance of the space should not be so small that the energy beam will have so small space to process or to function, or there will not be any space for the focal length to get its shape, or the space should not be so small that moving scraper will collide with the roof of the processing chamber, But if moving parts within the chamber can move well without any collision, the beam can be manipulated and the processing can be monitored, then the processing space is adequate. The processing chamber does not want any more room because in this minimum space, it can process a lot of parts, the number of parts can be million.

If the number of parts will increase too much, problems start to happen, but this problem is not because the processing chamber requires more space. After all, even if million parts are formed, it is not the job of the processing chamber to possess them after their fabrication. For possessing these parts, there is already another space, i.e., machine space or build volume. If million parts cannot be stored in the machine space, it is not because the processing space fails in its duty to transfer the parts layer by layer to the machine space but because the machine space is not big enough to be able to fulfill its duty to receive the parts layer by layer from the processing space.

It does not mean when the machine space will become bigger to store many parts, it will not affect the size of the processing space, it may affect the horizontal dimension of the processing space. It only means the machine space and processing space are two different spaces in BS, they are not linked together in the same way as these two spaces are linked together in deposition system (DS) (Fig. 8.6).

In DS, parts are formed layer by layer deposition in a processing chamber. As layer by layer deposition continues, the deposited layer has nowhere to go, the deposited layer does not slip down to give space for the next deposited layer to come and occupy (Fig. 8.6b). There is no space available for the deposited layer to slip down, there is no mechanism in place to facilitate the layer by layer vanishing of the part from the processing chamber. When the layer is deposited, the layer occupies the space in the processing chamber. When the next layer will be deposited, the next

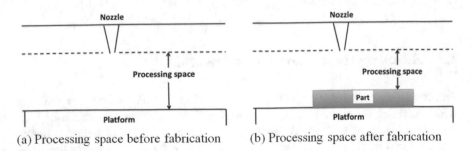

(a) Processing space before fabrication   (b) Processing space after fabrication

Fig. 8.6 Processing space in DS. (**a**) Processing space before fabrication. (**b**) Processing space after fabrication

layer does not have the same space as the preceding layer had. The preceding layer has already encroached the processing space resulting in the shrinking of processing space. Layer by layer encroachment continues until there is no more processing space available, or there is processing space available but that space is not sufficient to be able to do deposition well.

Layer by layer encroachment can be defeated by making the processing chamber without roof so that as the build moves layer by layer in the vertical direction, the nozzle moves up layer by layer to maintain the working distance that is already optimized for the right deposition. More the progressive layer moves up, more the nozzle moves up. Thus, the nozzle is no longer have limited space to move up. The nozzle no longer requires another parameter for deposition because the same parameter no longer works in a small space. The encroachment is defeated. When the encroachment is defeated, it reminds the same situation in BS where there is also a constant working distance. In BS, the constant working distance is maintained by pushing down the layer whenever a layer is finished while in DS, the constant working distance is maintained by pushing up the nozzle whenever a layer is finished.

In DS, an infinite number of parts can be formed if there is no roof available to restrict the movement of the nozzle when the nozzle is pushed up layer by layer while in BS, an infinite number of parts can be formed if there is no ground available to restrict the movement of the finished layer when the layer is pushed down layer by layer. It does not mean BS should be kept on Moon to make million of parts, there are solutions available which are given henceforth.

Figure 8.6a shows available space in DS. The available space shrinks with the fabrication of the part as shown in Fig. 8.6b.

Hence, in DS, the processing space is the machine space, the capacity of the processing chamber is the build volume of the machine. There is no concept of two spaces, i.e., machine space and processing space.

This is same in injection molding (IM) where there are no two spaces, i.e., one for fabrication and the other for the hoarding. In IM, after parts are fabricated in mold, they need to be ejected out from the mold to give way for another cycle to start. Thus the parts that are formed in the mold do not get stored somewhere in the mold, after the fabrication, which is in contrast to BS. The movement of parts in IM works on the basis of first ready first go while the same in BS works on the basis of ready anytime but go together.

1.7   Advantage of AM Over Injection Molding

For fabricating such large number, IM requires the fabrication of mold or tool first which will delay the production. Thus AM will provide fast fabrication of high number of parts (refer Chap. 1). But if fabrication of mold is faster (due to use of AM for mold fabrication) [9], IM is no longer slower for the reason that there is a delay in mold fabrication. Since AM is a slow process while IM is fast, AM then does not have advantage over IM in terms of time.

If AM is not faster than IM for production in mass scale, still AM can be a preferred choice for production because AM is a direct process. AM provides one-point solution. Using AM means there is no need to search one machine for mold fabrication and then search another machine (for part fabrication) that will fabricate employing that mold. Besides, there is no need to bring two different machines under one roof so that production will not be delayed because the second machine is still waiting for the mold to arrive and the production process to resume.

If AM is not faster than IM for production in mass scale, still AM can be a preferred choice for production because AM can be cost-effective. The cost of fabrication in BS is mainly driven by the machine time. Thus if one part is fabricated, its cost is higher than the cost of one part when many parts are fabricated. Though machine time increases with an increase in number of parts but the increment in machine time is less than if each part is fabricated in a different machine. Because if all parts are fabricated in the same machine, increment in machine time is mainly due to scanning time. Otherwise, increment is due to scanning time plus machine setting time.

1.8   Role of Machine Time

Thus in BS, if many parts are fabricated, the cost per part will become low. If many more parts are fabricated such as production is approaching to mass scale, the cost per part will become further low. Consequently, if a large number of parts is fabricated, the machine time will increase and the cost per part will decrease. Thus in BS, the machine time plays a central role to decrease the cost.

In IM, the machine time again plays a central role to decrease the cost but this role is not the same as in BS.

In BS, this is the machine time which governs how many of thousand parts which are already started to be fabricated will be finally fabricated. If there is small machine time, few out of total incomplete parts will be completed. If there is large machine time, many out of total incomplete parts will be completed. In other words, if there is small machine time, few rows out of all rows will be completed. If there is large machine time, many rows out of all rows (Figs. 8.1c, 8.4) will be completed. Therefore, in BS, machine time is related to the machine space.

For few machine time, few rows are occupied and therefore few machine space is in service. With an increase in machine time, more rows are occupied and therefore more machine space is in service. But, machine time cannot be arbitrarily increased if there is a limited machine space. There is a limitation after which there is no increase in production and there is no more decrease in the cost per part. Increasing the production then ceases to be a strategy to bring the cost down.

In IM, the role of machine time is different. If there is small machine time, few parts are initiated in fabrication and these few parts are completed. If there is large machine time, few parts are initiated in fabrication and these few parts are completed; after that, few parts are initiated in fabrication and these few parts are completed and so on. Finally, many (few + few and so on) parts are completed.

Therefore in IM, machine time is related less to machine space and more to its production rate. Machine time can be arbitrarily increased even if there is a limited machine space. The non-restriction in machine time leads to unlimited production, consequently unlimited decrease in cost per part. If the production is not unlimited, it is because the machine breaks down after some time or materials get exhausted after some time or some other issues crop in. But it is not unlimited because there is a limitation in machine space as happens in BS.

1.9   Effect of Machine Space

The difference in production quantity in BS and IM is not just due to the machine space and production rate but what happens in the processing space in BS. Material is processed longer in BS than in an IM machine for making a part. Thus AM is a slow process while IM is fast. Consequently, the same machine time produces a small number of parts in AM while a large number of parts in IM. The cost per part due to machine time is high in AM while is low in IM. But in IM, the cost per part is not only due to the machine time but due to the cost of the tool as well.

Consequently, if a single part is fabricated both in BS and IM, it is expensive in BS due to machine time but it is much more expensive in IM due to tooling cost. With an increase in number of parts, this trend will continue before the trend is reversed [10]. As long as the trend continues, AM or BP is more cost-effective than IM. This may be some hundred parts or some thousand parts [11] depending upon type of material, size of the part, tooling cost, complexity of the part (influencing machine time), etc.

1.10   Million Parts

If several thousand parts are produced in BS, it is itself an achievement when fabrication of few parts is a norm. Production of thousand parts will bring limitation to the system – the machine will be exhausted after several weeks of operation (refer Chap. 5), materials will be over.

But if the aim is to have hundred thousand parts, what will happen. Machine can be made bigger, but scaling up is already limited and this option is not open. Alternatively, several machines such as hundred machines will be procured. This option is not new and is presently done to increase the production in a limited scale. Moreover, taking help of several machines is not an improvement in production system. Several machines may not help production sustainable as well (refer Chap. 6).

When machine works for several weeks, it stops working because either build volume is full or all material is consumed. Machine starts working if it is started to further increase the production quantity. Machine will stop again for getting refilled. The cycle of starting and stopping must not go indefinitely, it cannot be a strategy to improve mass production, it is not because machine cannot be made more robust but because the machine with the present configuration is not meant for such continuous type production.

If machine is modified, many more parts can be produced. The problem with the machine is the cycle of starting and stopping. If there is no requirement for such cycle, production can flow continuously. The main reasons for the cycle are limited material and build capacity, but are minor reasons as well such as limited longevity of beam, weaker machine parts, lack of monitoring, etc. But even if these minor reasons are improved, there will be no change in the cycle while the main reasons, if solved, will lead to the production of many parts.

If AM machine is connected to a tank of materials which supply constant stream of materials, and these materials replenish without stopping the working of the machine, the problem of the materials is over. The build volume furnished by the downward movement of the platform is limited, the build volume cannot be made many times bigger. Thus even if there is no dearth of material in the running machine, the machine needs to be stopped to empty the build volume. Emptying the build volume takes time, by the time it will be emptied, it already consumes time for some layers or parts to be built up. The consumption of this time brings discontinuity in the production.

In order to maintain the continuity of the production, the moment the filled up platform leaves BS, other empty platform must come up. The movement of platform is not new (refer Chap. 9), it works in modular based AM machines (Concept Laser GmbH), but the use of modules in this case is not to increase product quantity but to increase product quality by erosion (to smooth each layer), a post-processing type action (refer Chap. 4).

1.11 Window of Time

The delay in the coming of empty platform will not be able to break the continuity of the production, it is not because the empty platform will take fraction of second to arrive and be fitted but because there is no need for the empty platform to come so fast. It takes time for the preparation of spreading the layer. The time the machine takes to prepare for spreading the layer, no fabrication goes on. Hence, the absence of fabrication during the spreading does not make the fabrication process discontinuous, because spreading the layer or related exercise for spreading the layer by stopping the layer consolidation (and hence part fabrication) is an essential part of BP. The uniqueness of the process gives a window of time in which the old platform can be exchanged with the new platform without disturbing the continuity of the fabrication (Fig. 8.7).

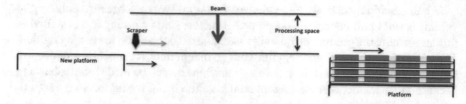

Fig. 8.7 Continuous flow of production in BS: old platform is leaving AM machine and new platform is coming

In IM, when the material is injected into mold, transformation of the material into part starts and when the part is ejected from the mold, the transformation ends. This transformation for a part is a continuous process. The transformation does not stop before the part is formed. The transformation can stop for some other reasons such as lack of sufficient materials or defects in the machine but it does not stop because stopping the transformation is one of the essential requirements of IM process.

In BP, the transformation of material into part does not happen continuously. The transformation stops after every layer for the next layer to deposit. But the stopping of the transformation after each layer does not make the process discontinuous because stopping the transformation after each layer is required to move to the next stage of the process.

Thus in IM, the transformation of material into part and continuity of the process is one or the same thing. If there is no transformation, the process will stop. While in BS, the transformation of material into part and continuity of the process are two things. If there is transformation, the process goes on. If there is no transformation, the process is still going on because some layer is getting spread.

When parts need to made in a large number, there should not be delay in the fabrication of the next part. If there is delay, production will be delayed. If the production is meant for a large number of parts, the delay even in 1 s will cause enormous delay for mass production. But when the delay is essential and the delay is an inherent aspect of the process, the production cannot be expedited by removing that particular delay.

If there is a delay in the production due to this particular delay and there is no other delay, there should not be any worry because the production is running at its maximum speed. There should not be any concern anymore because the production has achieved its maximum potential. If there should be any concern, it should not be about that particular delay but should be about other delays that occur due to myriad reasons such as lack of planning, lack of robust systems, etc.

In BP, the discontinuity in transformation due to delay in the layer spreading is due to an essential delay, the production cannot be expedited by removing this essential delay. This type of expediting the production is not expected as well. This is not a concern as well. But what is concern that even in the absence of such particular delay, the production is slow. But what is the bigger concern is not that the production is slow but the production will become slower. The production will become slower when mass production is desired. Because when mass production is desired, production needs to move from one batch to another batch and to another batch—this will continue. The delay in transferring from one batch to another batch will be accumulated. There is no mechanism in place to overcome such delay. Thus, essential delay plus other delays will make the mass production slower.

It is time to blame the process. The process needs to be less blamed for the reason why the production is slow because the delay due to layer spreading is an essential part of the process. The process needs to be more blamed for the reason why the mass production through it will be slower. The process provides the solution. The very reason why the production is slow is the same reason why the mass production will not be slower. If the time due to layer spreading is an essential delay, the same time due to layer spreading has potential to accelerate mass production if the same time is utilized. The same time will be utilized when moving from production of one batch to another batch does not take any time. It is not taking any time because the production is moving to the next batch only when the process is already consuming time doing its essential ritual, i.e., layer spreading.

Moving from one batch of production to another batch takes time even in IM or any other CM, but BP gives opportunity that henceforth moving from one batch of production to another batch will not take any time. What BP is able to give cannot be obtained from any CM. The very reason why BP is blamed is also the reason why BP can claim to be better than all CM engaged in batch production. This is the uniqueness of BP.

It can be argued that IM is faster than BP because IM does not have such uniqueness. If IM had such uniqueness, IM would not be faster. Therefore, what is the uniqueness of BP is limited only to BP.

The point is not how BP equipped with such uniqueness will become better than IM or any other CM. The point is how BP will perform better in the present circumstances. Though, BP can have many problems, but it has some aspects which if can be utilized, it will perform better. In order to know those aspects, it is essential to compare those aspects with other processes. If one of those aspects is the best and can be termed uniqueness, it is not exaggeration in the light that other processes (IM, etc.) do not have such aspect. It does not mean other processes should try to possess such uniqueness. It only means if that uniqueness of BP will not be recognized, BP will not perform better.

1.12   Another Option for Continuous Production

If there is requirement for million parts, there is no one new platform replacing an old platform but many platforms need to be used. Figure 8.8 shows the continuous supply of materials and platforms to an AM system. In this case, BS will not be an isolated system, which is the case these days. The system should be ready to handle inputs and to transfer products. It means BS will be a small part in a big automated

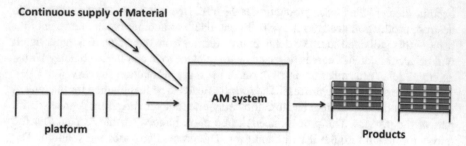

Fig. 8.8 Schematic diagram for the continuous flow of production in BS

network of production line—this is what can also be known as direct digital manu-facturing (refer Chap. 9).

This brings a question what can be an alternative for using so many platforms. There is a solution. It is to use only one platform instead of many platforms, so that when it is full, it can be emptied fast. Thereafter, the empty platform can then be refit-ted in the AM processing chamber within the layer spreading time (window of time).

Emptying the build volume or platform (Fig. 8.9a), after the fabrication, takes time because it is not configured for emptying fast and reusing. Imagining a build chamber which has a sliding door so that after the fabrication, when the platform is tilted, the parts slide away (Fig. 8.9b). The platform then gets back to its original position (Fig. 8.9c) ready to resume the fabrication.

Parts can slide away if they are not bonded to the substrate placed on the plat-form. If they are bonded, substrate or build plate needs to be removed as well along with parts. If the substrate is removed, another substrate is required. Thus continu-ous production means continuous supply of substrates or build plates and continu-ous fitting of the substrate on the platform within the window of time.

Reusing the platform for continuous production does not mean being free from the continuous supply of an essential part (or section) of the platform. When there was the continuous supply of platform, there was no freedom from the new plat-forms and when there is no continuous supply of platform, there is no freedom from the new parts (of the platforms) as well. Bigger problem, i.e., supply of bigger entity (platform) is replaced with smaller problem, i.e., supply of smaller entity (build part of the platform), the problem is not gone completely.

But in some cases, the problem will go and freedom can be achieved. When the part is not bonded to the platform, this is mostly in the case of polymer powder bed fusion, all parts can be removed without detaching the parts (without breaking the bonds) from the platform or the substrate. There is then no need of new substrates for continuous production, there will be need for removing the parts from the sub-strate. It is new in BS, but limited automatic removal of parts using robotic arm is reported in extrusion based AM [12]. Removing the parts fast by sliding may break some fine features of the parts and will not be convenient for delicate parts. Thus this option is limited for those parts which are not bonded to the substrate and will not get damaged during removal.

Fig. 8.9 Emptying the
platform to reuse it: (**a**) BS
after fabrication, (**b**)
Emptying the platform by
tilting, (**c**) Position of the
platform after emptying

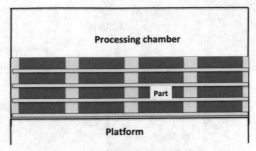

(a) Parts in BS after the completion of
fabrication

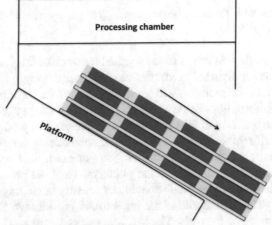

(b) Emptying the build chamber

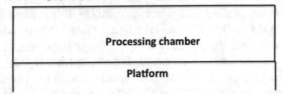

(c) Platform in its original position after
emptying the parts.

1.13   Metal and Polymer Systems

BS is of three types, i.e., powder based, slurry based, and photopolymer based.
Since photopolymer based BS mostly works in a vat (container), and the fabricated
part slips in the vat (Fig. 8.10). The movement of platform as a means to expedite
the production quantity will not work with a vat.

The vat is full of liquid. The vat is the build space. The maximum size of the part
or maximum volume of parts is limited by the size of the vat. The vat plays double

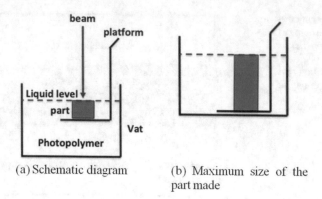

(a) Schematic diagram   (b) Maximum size of the
                                           part made

Fig. 8.10 Photopolymer bed system (PPBS): (a) common system configuration, (b) fabricated part in vat

role. Its first role is to supply the feedstock to the platform. To supply the feedstock (photopolymer liquid), the vat does not do anything, it remains at it is and waits when the platform will go down. When the platform goes down, the level of the platform goes below the level of liquid of the vat (Fig. 8.10a). This submerging of the platform in the liquid or this submerging of platform in the feedstock allows the feedstock to be on the platform. If submerging does not help feedstock to be on the platform, blade is used to bring it. For example, blade is used to push high-viscous liquid from the side of the platform. Thus vat plays its first role of supplying the feedstock to the platform without actively supplying the feedstock to the platform.

The second role of the vat is to act as a storage of the parts. The platform is the storage of the parts. The platform can continue storing the parts until it touches the base of the vat. Thus the parts are in the vat placed in the platform (Fig. 8.10b). Replacing the platform means replacing the platform through the vat. It means moving the platform through the liquid. This is the problem. But the problem was not there when the platform needed to move in powder bed system (PBS).

In PBS, there is no vat to play double role. In PBS, there is a build space to store the part, and there is separate powder container. The downward movement of build space does not affect the powder container. The downward movement does not cause powder to appear on its own on the platform because powder is not liquid. Metallic powder gives stronger part than photopolymer liquid can, and is thus better than liquid when a stronger part is desired. But it gives problem that it cannot appear on its own on the platform when the platform is pushed downward.

This brings limitation to the type of system that can be configured using powder. When powder has problem that it cannot appear on its own, it requires to be carried to the platform. When it always requires to be carried, then only that system can be configured which allows space for such carrying. But liquid is free from such limitation, it does not require always to be carried. Liquid can appear on its own even if there is no space for such carrying. Thus, using liquid, those systems can be configured that cannot be configured with powder.

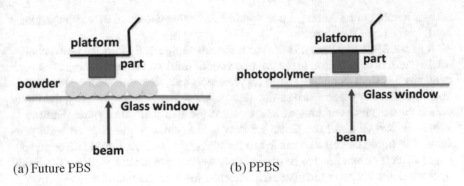

Fig. 8.11 Upside down system configuration: (**a**) future powder bed system, (**b**) photopolymer bed system

What will happen if PBS is configured upside down. It will bring many questions. How energy beam will strike the powder on the platform. How powder will be carried to the platform? It is possible to configure that system as well. Energy beam will strike from the below on the powder kept above it on a glass platform or a glass window (Fig. 8.11a). The powder is supposed to be consolidated among themselves but not with the glass window. If, by mistake, they get attached to the glass, they should be easily detached when the part moves up leaving the glass free for the next round of powder coming. When the powder is carried with the help of a scraper on the glass, the part moves up again to give space for the scraper. When the scraper goes back after depositing the layer, the part moves down touching the powder layer so that when the energy beam will strike again, the layer will be consolidated with the part, and the part becomes bigger by one solid layer.

Thus, a solution is found. Moving part up to accommodate the height of the scraper, and moving down the part when the scraper went back after depositing the layer. This moving up and down is a solution. This means there is no need to worry about non-flowing of powder like liquid. But, this is not a solution, this is a big problem. Moving up and down only because there is a scraper means moving up and down for some non-essential job. Moving up and down takes time. If there is mass production aimed, this time taking will be costly. Moving up and down means doing exercise against gravity. There is no more luxury left when the part had only to move down along with the platform in the common configuration. At that time, the system was not strained because the platform had not to periodically move up as well after every layer.

When the movement demands a precision of some microns, the system needs to be robust, and when the system needs to work against gravity frequently, the system needs to be more robust. When the system is expected to make thousands of parts, the system needs to be extra robust not only because the system needs to move up and down but the system needs to move carrying several kilograms weight. If the powder was low-viscous liquid, moving up and down only to accommodate the scraper would not be there; the part would have only concern to move up a distance equal to the layer thickness.

A new PBS is thus ready. But what is the advantage of that upside down configured system in comparison to the common configuration. Presently, there is no such configured system available. While for photopolymer, there are such systems available, where energy beam strikes the glass window from the below [13], the liquid comes on the glass window, which is one of the two platforms; other platform is above the part (Fig. 8.11b). Glass window is also called build plate or build platform. The liquid can also be carried to the glass window, but liquid unlike powder does not have compulsion to be carried only to give rise to such system. The liquid does not get attached to the glass, it is attached more to the platform above it.

When the platform moves up, the solid layer remains attached with the platform while it remains detached from the glass. To make the bond weaker on the glass and facilitate the detachment, Teflon coating is applied on the glass. It causes the bond between the solid layer and coating to be weaker. Another way to make bond weaker is not to allow any bonding to take place in the vicinity of the glass. If there will be no bond, question will never arise whether bond is weak or strong. In order not to have bonding, photopolymer needs to get solidified everywhere except just where it touches the glass. It can happen if oxygen is diffused through the glass [14], which inhibits solidification near the glass and saves the glass from the cycle of attachment and detachment.

This type of periodic attachment and detachment is not possible with metal powder. It is not expected when the metal powder layer will be solidified on the glass, it will be detached without breaking the glass only because it needs to go up being attached with the platform. This is the reason why there is no metal PBS available in upside down configuration. But what if metal powder is replaced with polymer powder. Solidified polymer layer is not so strong to break the glass window. A solidified polymer layer made from polymer powder is not enough different in properties from a solidified photopolymer layer made from photopolymer liquid to stifle the development of polymer powder BS if it is planned to develop.

But, when a polymer PBS is planned to be developed in upside down configuration in future (Fig. 8.11a), before it needs to be asked why that type of system needs to be developed. Whether it only needs to be developed to prove that whatever the type of systems can be achieved with photopolymer can also be achieved with polymer powder.

If there is PPBS available in that configuration, there is a reason for it. The reason is that this configuration makes vat free from playing double role. The part henceforth made will be free from getting submerged in the vat. The part is henceforth free not to be damaged by the presence of liquid if it was getting damaged in the past. The part is henceforth free not to break any of its features if it broke when it was getting submerged layerwise in the past. The vat is also free from for not allowing its liquid to be damaged by the beam. In common system configuration, when the beam scans the layer, the effect of the beam does not remain confined only to the layer, but it reaches to other areas of the vat, causing the resin to age prematurely [15].

If PBS is developed in upside down configuration, there will not be relative advantage because PBS unlike the PPBS is not affected by any such container. Thus, if it is developed, it needs to be developed for some other reasons.

If PPBS is in upside configuration (Fig. 8.11b), it will facilitate movement of new platforms without going through vat. Thus, if mass production is achieved in PBS in common system configuration, it needs to be achieved in PPBS in upside down system configuration because it will not be achieved in PPBS in common system configuration. Upside down configuration is called bottom-up while common configuration is called top-down [16–18].

1.14 Processing of Metal and Polymer

For consolidating polymer, small energy through beam is required. Thus a single beam can be divided into several beams using mirror based scanning system, such as DMD. Thus, in the case of polymers, simultaneous scanning of all parts is possible which will increase the mass production rate. In the case of metal powders, high energy beam is required, and there is a limited option for scanning. High energy beam is not suitable for low strength mirror of DMD. However, Texas Instruments is developing new DMD which can use near-infrared laser of power up to 160 W, suitable for processing polymer powder. In the absence of DMD for metal, mass production will be faster for photopolymer.

When high energy beam is used to melt metal powders, heat is generated in the chamber. With an increase in a number of parts, the heat accumulated will be high enough to change the part properties. In order to avoid this, heat needs to be removed from the chamber. The heat can be removed by pausing the processing after some layers, or increasing the idle time between two sets of layers. But, this decreases the production rate, and is not suitable when mass production is aimed. Another option is to cool by convective heat transfer using airflow [19]. This is effective when air flow does not happen far away from the processing zone or layer going to be processed. But, if the air flow is too close, it will change the part properties by cooling melt pool, or it will change the composition of the part by blowing some of vapor or plume generated during melting.

In the case of photopolymer, low energy beam is used and is therefore no need to manage heat. But, processing resin creates fumes which need to be vented out [15].

1.15 Comparison with CM

Though continuous production using an AM system can cater million parts but since AM is slow, this continuous production will be slow continuous production. Million parts can be obtained, if not after several years then at least after several months of continuous operation while mass production in IM takes several weeks.

This brings a question on the investment to convert an AM system into a production system.

Due to its unique capability, AM is already moving into the direction of many parts production. Since AM system is not presently capable for continuous flow production, other non-sustainable methods (using several AM machines) are getting practiced. Converting AM system into production system will not help only make environment sustainable but also make AM sustainable. Though AM is slow but when unique parts are made, AM becomes fast. It is not because basic characteristic of AM changes but because CM becomes slower (refer Chap. 3). Producing such million parts, even if takes several years, is not slow when seen from the perspectives of CM.

References

1. Wiese, M., Thiede, S., & Herrmann, C. (2020). Rapid manufacturing of automotive polymer series parts: A systematic review of processes, materials and challenges. *Additive Manufacturing, 36*, 101582.
2. Wiese, M., Leiden, A., Rogall, C., et al. (2021). Modeling energy and resource use in additive manufacturing of automotive series parts with multi-jet fusion and selective laser sintering. *Procedia CIRP, 98*, 358–363.
3. Deradjat, D., & Minshall, T. (2018). Decision trees for implementing rapid manufacturing for mass customisation. *CIRP Journal of Manufacturing Science and Technology, 23*, 156–171.
4. Roy, N. K., Behera, D., Dibua, O. G., et al. (2019). A novel microscale selective laser sintering (µ-SLS) process for the fabrication of microelectronic parts. *Microsystems & Nanoengineering, 5*, 64.
5. Lambert, PM., Campaigne, EA., & Williams, CB. (2013). Design considerations for mask projection microstereolithography systems. In *SFF Proceedings* (pp. 111–130).
6. Tenbrock, C., Kelliger, T., Praetzsch, N., et al. (2021). Effect of laser-plume interaction on part quality in multi-scanner laser powder bed fusion. *Additive Manufacturing, 38*, 101810.
7. Zhang, W., Hou, W., Deike, L., & Arnold, C. B. (2020). Using a dual-laser system to create periodic coalescence in laser powder bed fusion. *Acta Materialia, 201*, 14–22.
8. Sow, M. C., Terris, T. D., Castelnau, O., et al. (2020). Influence of beam diameter on laser powder bed fusion (L-PBF) process. *Additive Manufacturing, 36*, 101532.
9. Huang, R., Riddle, M. E., Graziano, D., et al. (2017). Environmental and economic implications of distributed additive manufacturing: The case of injection Mold tooling. *Journal of Industrial Ecology, 21*, S130–S143.
10. Chen, D., Heyer, S., Ibbotson, S., et al. (2015). Direct digital manufacturing: Definition, evolution, and sustainability implications. *Journal of Cleaner Production, 107*, 615–625.
11. Achillas, C., Aidonis, D., Iakovou, E., et al. (2015). A methodological framework for the inclusion of modern additive manufacturing into the production portfolio of a focused factory. *Journal of Manufacturing Systems, 37*(1), 328–339.
12. Aroca, R. V., Ventura, C. E. H., Mello, I. D., & Pazelli, T. F. P. A. T. (2017). Sequential additive manufacturing: Automatic manipulation of 3D printed parts. *Rapid Prototyping Journal, 23*(4), 653–659.
13. Gurr, M., & Mülhaupt, R. (2012). Rapid prototyping. In K. Matyjaszewski & M. Möller (Eds.), *Polymer science: A comprehensive reference* (pp. 77–99). Amsterdam: Elsevier.
14. Januszewicz, R., Tumbleston, J. R., Quintanilla, A. L., et al. (2016). Layerless fabrication with continuous liquid interface production. *PNAS, 11*(42), 11703–11708.

15. Salonitis, K. (2014). Stereolithography. In S. Hashmi, G. F. Batalha, C. J. Van Tyne, & B. Yilbas (Eds.), *Comprehensive materials processing* (pp. 19–67). Amsterdam: Elsevier.
16. Hafkamp, T., Baars, G. V., Jager, B. D., & Etman, P. (2017). A trade-off analysis of recoating methods for vat photopolymerization of ceramics. *SFF Proceedings, 28,* 687–711.
17. Santoliquido, O., Colombo, P., & Ortona, A. (2019). Additive manufacturing of ceramic components by digital light processing: A comparison between the "bottom-up" and the "top-down" approaches. *Journal of the European Ceramic Society, 39*(6), 2140–2148.
18. Behera, D., Chizari, S., Shaw, L. A., et al. (2021). Current challenges and potential directions towards precision microscale additive manufacturing – Part II: Laser-based curing, heating, and trapping processes. *Precision Engineering, 68,* 301–318.
19. Ho, J. Y., Wong, K. K., Leong, K. C., & Wong, T. N. (2017). Convective heat transfer performance of airfoil heat sinks fabricated by selective laser melting. *International Journal of Thermal Sciences, 114,* 213–228.

Chapter 9
Future

1 What Will Be the Concept of Direct Digital Manufacturing in Future?

1.1 In Search of Direct Digital Manufacturing

Earlier, rapid prototyping or additive manufacturing (AM) used to be called direct digital manufacturing (DDM) [1]. The meaning of DDM was a type of manufacturing that converts a digital file directly into a product. The meaning was not different from the meaning of AM. AM exactly does the same, thus the term DDM was redundant in the same way as the terms rapid manufacturing, rapid tooling, rapid prototyping, 3D printing were redundant (refer Chap. 1).

The use of DDM then was more to appreciate and emphasize the digital aspect of AM. It had no practical extra meaning, it was not conveying a new approach to manufacturing other than what AM already was conveying. Its use was more to convey the promise of AM when AM was trying to get acceptance in industries.

The meaning of DDM evolved with the advancement of AM. Henceforth, DDM meant a method which is such type of AM which allows many products to be formed. Since AM can design such many product in many ways, DDM meant a method for mass customization [2]. But many AM processes were not able to fulfill this criteria, DDM inadvertently represented few AM processes. The meaning of DDM in this case again was not different from the meaning of AM. This meaning was more to advertise the production capability of AM rather than initiating a new manufacturing paradigm. Thus, DDM was again redundant.

Then, DDM had nothing to do with the automation. If AM as DDM was expected to do mass customization, it meant only that an AM system would make a number of customized products. The meaning of DDM did not include how materials will be automatically transferred to an AM machine, or how products made from an AM machine will be automatically collected—these were not the expectations from

© The Author(s), under exclusive license to Springer Nature Switzerland AG 2022
S. Kumar, *Additive Manufacturing Solutions*,
https://doi.org/10.1007/978-3-030-80783-2_9

AM. If AM system is making a number of products, it is already fulfilling the criteria of DDM. The automation only implied that how an isolated AM system functioned well without the need for manual input, or how various components of the isolated system were automated without having any connections with peripheral units or cyber-physical systems. This automation was the minimal automation. Not emphasizing automation beyond an isolated AM system was also for a practical reason—an accepted product needs to be formed first before an extended automation is sought.

AM was growing, DDM had to grow as well. DDM started to acquire new meaning. Henceforth, if AM system can be connected to the internet and a product can be formed by a distant customer, this can be DDM [3]. But connection with the internet and many other systems is what digital manufacturing is. DDM was no longer equal to the direct conversion of a digital file to a product, but was equal to the direct conversion of a file plus digital manufacturing.

The direct conversion to product is not the privilege of only AM, other conventional manufacturing (CM) processes such as machining do it as well. The direct conversion means AM does not need to make a mold for making a product while mold based CM (such as injection molding, die casting, sintering, etc.) needs to make a mold for making a product. Thus, the direct conversion takes place in AM while direct conversion does not take place in mold based CM. Consequently, AM is better than mold based CM in those situations where direct conversion is a virtue (refer Chap. 1).

But, AM does not have any advantage over non-mold based CM (such as machining, drilling, milling, etc.) which are also capable to do direct conversion and do not require any mold to carry out conversion. If AM has advantage over non-mold based CM, it is because AM can make some complex products which non-mold based CM cannot make. This advantage is not because non-mold based CM lacks the quality of direct conversion but because non-mold based CM are different. Being different from AM does not make CM less efficient in all cases (refer Chap. 3).

It can be argued that non-mold based CM though do not require mold, they still require design-specific tooling while AM does not require design-specific tooling. Therefore, non-mold based CM are unable to perform direct conversion of the file while AM does direct conversion. Consequently, AM is better than non-mold based CM. This inference is based on the fact what will happen when a complex product needs to be formed. When a complex product is required, non-mold based CM require design-specific tooling while AM does not require design-specific tooling, therefore non-mold based lacks the quality of direct conversion while AM does not lack.

Thus, this inference takes help of complexity to prove why non-mold based CM lacks the quality. This inference will not hold good if the complexity does not arrive. The complexity arrives at the next stage. Before the next stage, non-mold based CM can make a number of products without requiring more than one design-specific tool. If a non-mold based CM is using only one design-specific tool, this process is free from the obstacle of having more design-specific tools to enable direct conversion of the file. The quality of AM is free from the obstacle of having design-specific tools. When a non-mold based CM is using only one design-specific tool, it possesses the same quality what AM possesses. But, a non-mold based CM lacks the quality when complexity arrives. What happens at the next stage cannot deny what happened before the next stage. Similarly, what will happen at the advanced stage cannot undo what happened at the earlier stage. Both processes non-mold based CM and AM are basically same that cannot be negated because some complex examples give the impression otherwise.

It can be argued what happens at the earlier stage is not important, but what happens at the advanced stage is more important because the aim is to make complex products, the aim is not to make a simple product and that is why what happens for the fabrication of a simple product is not important. This argument is bigger excuse than the earlier excuse to rely always on the complexity-card to evade the basic issue with an aim to prove the supremacy of AM over CM. It is not denying that the complexity is important and therefore a process has to prove its efficiency by making complex products. But, the problem starts because AM does not always succeed to make complex products while CM does not always fail to make complex products (refer Chap. 5).

Thus, as far as direct conversion is concerned, both AM and non-mold based CM stand on equal footing. Some advantage of AM over non-mold based CM and vice versa does not change the fact that both AM and non-mold based CM equally possess the basic characteristic.

Hence, DDM should be equal to digital manufacturing plus only AM was not a logical conclusion. If it was digital manufacturing plus any manufacturing process, it would be more logical. DDM then became digital manufacturing plus any manufacturing process albeit with some conditions [4]. It did not mean the evolution of DDM had been leading to any consensus on the use of DDM. As a result, it was again used as its oldest version, i.e., AM [5, 6].

In order to have a meaning of DDM, its meaning in an ideal case in relation to AM is given—if AM is fully developed in future, then what are the conditions that AM needs to fulfill to be called DDM.

Ideal case again implies a condition which may not necessarily improve the production process and efficiency but which will help know the difference between digital manufacturing (DM) and DDM. For example, it is henceforth assumed that outside intervention or manual intervention is not required for DM or DDM to take place. If there is any manual intervention, the ideal condition of DM or DDM gets disturbed. Need for manual intervention inadvertently implies that the automation is yet not complete. In reality, it is not always beneficial to remove every trace of manual contribution. For many cases, on the contrary, manual contribution is required to improve production efficiency. Nevertheless, this fact is ignored henceforth as the purpose is to convey the difference between DDM and DM.

1.2   Digital Manufacturing

DM is a state of interconnected manufacturing systems which enables digital file
and material to move, without human intervention or outside assistance, from the
first process to the last stage and convert the file into a physical product (Fig. 9.1).
The to and fro movement of information due to interconnection helps the design of
the product to be reconsidered and improved [7].

Figure 9.1a shows an example of DM where a product is formed when material
and digital file go from Process 1 to Process 2, which are connected digitally such
as by computer or cloud server [8] so that the movement of information and material
is automated, which can take place without outside assistance.

Figure 9.1b is another example of DM where two different files are processed by
two different processes. The final product is an assembled product using products
from both processes. This example shows that there is no limit to the number of
processes, files, and systems that can comprise digital manufacturing as long as
there is no loss of digital connection among them warranting human intervention.

1.3   Direct Digital Manufacturing

Figure 9.2 is the concept of DDM where there is one process, one file, and one system. The product is formed when one digital file is processed in a system, the product is the physical representation of the digital file. The movement of information
[9, 10] and material is automated.

If fabrication of a product requires more than one process or more than one file
or more than one system such as shown in Fig. 9.1, DDM is DM.

Fig. 9.1 Schematic
diagram: (**a**) digital
manufacturing using one
file, (**b**) digital
manufacturing using two
files

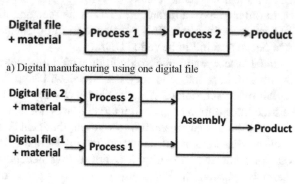

a) Digital manufacturing using one digital file

Fig. 9.2 Schematic
diagram of direct digital
manufacturing

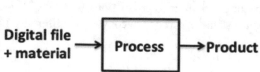

DM is for making products using connection, it does not matter how many processes or systems are required. But DDM is a type of DM, where digital connection is no longer as important as in DM. In an ideal case, no digital connection is required in DDM because there are not any units to connect to. It does not mean automation has no business in DDM, but automation is not for such lengthy outside connection excepting the internet, it plays role to automate a process within the system. In DDM, it is expected that one process should do everything what is usually done by a number of processes in DM. If one process is taking the responsibility of many processes, not only is the process well developed but also the automation is developed enough to enable the process. Automation is prerequisite if large number of parts need to be formed (refer Chap. 8).

If an assembled product is required, there are two processes and two digital files are required in DM but DDM makes such assembled product using one file and one process. So there is no need to establish a digital connection between processes and an assembly system because the process in DDM is developed enough to be free from such need.

1.4   Fabrication by AM

Figure 9.3 demonstrates that not every method to manufacture using AM is DDM. If AM uses a separate post-process system to finalize a product, AM is not DDM but can and cannot be DM. For example, a product is made in photopolymer bed system (PPBS), but the product property is not adequate, therefore the product is post-processed by curing in a separate oven. In this example, AM is neither DDM nor DM (Fig. 9.3) because there is no digital connection between AM and post-process systems.

If both systems are digitally connected so that production is initiated in PPBS and ends in a curing oven without outside intervention, AM is DM (Fig. 9.4). But AM is not DDM because two systems (instead of one system) are used.

If, for example, PPBS is advanced enough so that post-processing curing is no longer required in a separate curing oven but is done with the AM system, AM is DM but not DDM (Fig. 9.5). Though fabrication of the final product using photopolymer bed process did not require two separate systems but the product is not formed by AM process alone. The product is the result of AM process followed by a post-process (curing), thus there are two processes involved to make this product

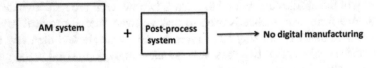

Fig. 9.3 No DM because of no connection between AM system and post-process system

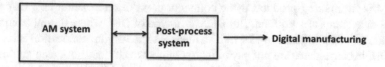

Fig. 9.4 DM because of connection between two systems but not DDM because of presence of two systems

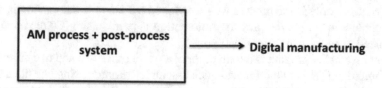

Fig. 9.5 DM because of one system confining both process and post-process but not DDM because of presence of two processes (one process and one post-process)

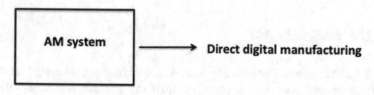

Fig. 9.6 DDM because of confinement of process within one system and no need of post-process

(refer Chap. 4). For AM to be DDM, AM is expected to be developed enough that it can make a product without the help of post-process.

If PPBS makes a product which is robust enough that it does not require post-process curing, there is no need for a curing oven, there is also no need for developing the AM system so that it can perform post-process, AM is DDM (Fig. 9.6).

1.5 Two AM-System

What if two-AM processes are confined in a system. Figure 9.7a shows two-AM processes, i.e., material jetting and material extrusion are connected in a gantry based system. For making a product using this system, a single digital file is required which generates tool path for both nozzles (one for jetting and other for extrusion) so that any of the nozzle can be used to make a two-material product. Two different materials one from each nozzle is used to make a two-material product. A simple product can be made by making few layers from one nozzle and then few layers from another nozzle so that two materials are alternately comprised in the product (Fig. 9.7b) (refer Chap. 7).

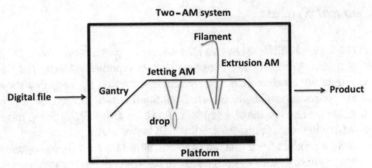

(a)Two AM processes (jetting based AM and extrusion based AM) in one AM system

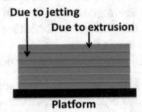

(b) Two-material product formed due to one system having two AM

Fig. 9.7 Two-AM processes (jetting based AM and extrusion based AM) in one AM system. (**a**) Two-AM processes (jetting based AM and extrusion based AM) in one AM system. (**b**) Two-material product formed due to one system having two-AM processes

This brings a question whether this system fulfills the criteria of DDM because there is direct conversion of a single digital file using one system. Although two separate AM processes are used but these processes are within one system. Incorporation of two processes did not warrant a connection between two different processes because these two separate processes were not confined in two different systems within a big system.

In this example, there are two different processes instead of one process, thus the system does not belong to DDM. Though the absence of difficulty in tool path programming and similarity between two different processes tend to believe the fact the system is using two separate processes. In spite of the operational similarity between two processes and commonality of nozzles among them, they belong to two different sub-categories of AM—one is solid deposition process (DP) while other is liquid DP [11].

Since there is no loss of connection during the movement of digital file from the beginning to the end for the fabrication of a product, the system (Fig. 9.7a) is a DM system.

1.6   Two AM Systems

If this is DM, how it is different from a DM when one process uses one system to be confined, thus two processes are confined in two separate systems and then two separate systems are connected. A new system is formed (Fig. 9.8a) that shows how two separate processes which were earlier confined within one system can be kept in two separate systems to make a product. Products (Figs. 9.7b, 9.8b) made from two types of arrangement (Figs. 9.7a, 9.8a) will be the same.

For making a product from this new system, some layers will be deposited on a platform from system 1, then system needs to be stopped. Afterwards, the platform needs to move to system 2 so that next layer will be deposited on it. When the platform reaches system 2, the platform is fitted in the system and system 2 will start the deposition; the platform will again go back to system 1 after the deposition. This back and forth movement of the platform will continue several times until the last layer is deposited and the product is formed.

Using two AM systems will give the following concerns:

- Assuming that the back and forth movement does not cause undue changes in the material state of the deposited layers. Otherwise, few more conditions need to be maintained so that the products by the new system will be exactly same to that from two AM system. If temperature of the platform changes during the movement, the platform must not need to move far. If environment affects the

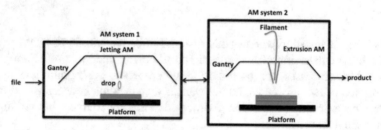

(a)Two AM systems: jetting based AM (AM system 1) and extrusion based AM system (AM system 2) digitally connected

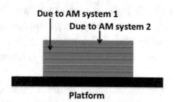

(b) Two-material product formed from digitally connected two AM systems

Fig. 9.8 (**a**) Two AM systems digitally connected (jetting based AM and extrusion based AM) in one system. (**b**) Two-material product formed from digitally connected two AM systems

deposited layer, the problem is to ensure not only how platform moves well but also how platform remains inert during the movement. It can be assumed that materials do not get oxidized during deposition. Otherwise, after the reattachment of the platform, the processing chamber needs to be purged with inert gas before the deposition.

- If inert environment needs to be maintained in the chamber after the reconnection of the platform, the deposition does not just start but has to wait until the right environment is maintained. Maintaining the environment several times in both systems, even if does not affect the product quality, can take considerable time. Assuming that the two systems are not near to each other but are located at the two ends of a factory, the movement of the platform needs to be scheduled in advance considering other movements. The platform needs to be robust enough that they can move and they should be intelligent enough to change their path in the anticipation of any collision with other vehicles. Though there are many problems when two separate systems need to be connected, but if thousands of products need to made in a factory setting where there is a long term benefit in multi-tasking using multiple systems, there is merit in overcoming these problems.

In spite of these problems associated with the movement, maintenance of systems during connection and reconnection, this new system (Fig. 9.8a) is an example of DM while in spite of avoidance of such problems, two AM system (Fig. 9.7a) is an cxample of DM as well. Avoidance of problems and improvement of system do not imply that the status of DDM is achieved. It is expected that an AM process is developed enough in future so that it can furnish, for example, a multi-material product without the help of another AM process.

1.7   Two-Material AM System

Figure 9.9a shows an example of AM system or DDM system where there is only one process but two-material products can be formed. It brings a question which two-material product is better—one that is made by using two processes in one system (Fig. 9.7b) or that one made by using one process in one system (Fig. 9.9b). Since both products are made differently therefore any one of them can be better. It is possible that a DM system (Fig. 9.7a) makes better product than a DDM system (Fig. 9.9a). Making a better product does not mean DM system will be elevated to get the status of DDM system while DDM system should be downgraded to DM system.

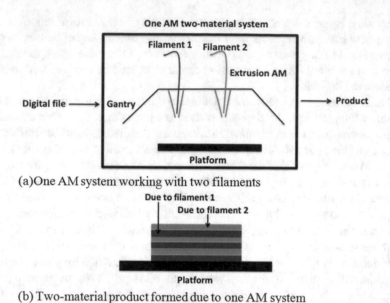

(a)One AM system working with two filaments

(b) Two-material product formed due to one AM system

Fig. 9.9 An extrusion based AM system to make two-material product. (**a**) One AM system working with two filaments. (**b**) Two-material product formed due to one AM system

1.8   Product Fabrication as a Criteria

The better product formation does not change the status of DM and DDM. But what if DDM system (Fig. 9.9a) is not able to make a product, the system is not a DDM system. Moreover, the system is not a system at all. It is the product formation that confirms the status of a system. But what if the DDM system (Fig. 9.9a) makes some products while it is not able to make some other products. If the DDM system makes some product, it confirms the status of being DDM. If the system is not able to make some other product, the failure in those product formation does not change the status the system of being DDM. But if the system tries to make only those products in which the system has always failed, the system has never a history to confirm its status of being DDM or otherwise.

1.9   Drilling

Thus, whichever manufacturing processes fulfill the product fabrication criteria is DDM. If drilling process is selected and if there is a product required such as a steel cube comprising of some holes, drilling is DDM. Since, drilling fulfills the criteria of one process, one file, one system and no outside assistance when product formation starts from the digital file and ends in the product [12].

If the product becomes complex such as there requires sizing of the cube to the half which drilling cannot do, another process (milling) is required. Hence, for a simple product drilling is DDM, but for a complex product drilling is not DDM. Complexity in a product determines to which extent the criteria of DDM is fulfilled.

It is argued that AM is a better process because AM makes the product layerwise while others do not, implying that layerwise manufacturing is itself an evidence that AM will make better products in future while other processes do not have such uniqueness. But AM is itself a generic process, there are many processes which are AM and all have different capabilities. There is powder bed fusion (PBF) [13] as AM which can make complex products while there is also wire arc additive manufacturing (WAAM) [14] as AM which is not able to make such complex products.

It does not mean drilling will succeed if fair chance be given to it and will outsmart PBF. It implies if the efficiency of a process is not a criteria, AM process having limited efficiency will be preferred to be the part of the latest manufacturing paradigm while efficient CM such as machining as direct digital subtractive manufacturing [15] or digitally controlled hose-cutting [16] will be excluded.

1.10   Injection Molding

All processes are not amenable or conducive to automation. For example, injection molding (IM) is neither DDM nor a candidate to reach DDM stage even if it is fully automated in future.

In IM, many identical products are formed using one mold. If different products need to be formed, the same mold will not work, other molds are required. Fabrication of different products depends upon the fabrication of different molds while fabrication of different molds depends upon conversion of different digital files into different molds (refer Chap. 1).

Assuming there is complete automation developed so that a digital file is converted into a mold by machining and the mold is fitted automatically into IM machine and the machine works automatically so that products are formed (Fig. 9.10). Alternatively, mold can be made by AM to accelerate the production process [17] or decreasing the cost of the mold [18].

After automation, if a new product or many copies of a new product is required, the digital file changes and a new mold is formed which replaces the old mold automatically and the products are formed. Thus starting from a digital file, a product can be formed at the end using IM without any human intervention anywhere from the file to the product.

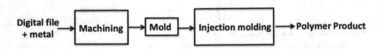

Fig. 9.10 Digital connection between machining and injection molding

But the final product that is formed is not a result of direct conversion of a digital file but the result of the utilization of an intermediate machine (i.e., IM), which is an intermediate step. The final product is made up of a different material that is not the material used when the digital file was converted into a mold. The final product is not having the shape and forms similar to that of the digital file. The direct digital conversion ends when the file becomes a mold. In the absence of direct conversion of file, IM is not similar to DDM. Though investing in automation and converting the digital file into a product, IM is fit to be digital manufacturing but not to be DDM.

Some automation is reported but it is mostly about automatic collection of products [19].

1.11 Industry 5.0

If different units shown in Fig. 9.1 are executed at different manufacturing systems separated by a distance, material handling from one system to another is required. In the case of a large scale production, factory floor plan is required so that material transfer should take minimum time and distance. Movement of material or unfinished product will require sensor based conveyance vehicle to avoid collision.

This gives rise to the concept of smart factory where every equipment is connected. The connection goes beyond the boundary of the factory and extends to the outer world. This means a work order generated by smartphone [8] far away from the factory is enough to begin and end production in the factory. This is the contribution of digital manufacturing for ongoing fourth industrial revolution. What will happen next—if there is fifth industrial revolution, what will be further advancement in the functioning of the smart factory.

The size of the smart factory will increase if the number of processes increases for making products. The smart factory will then become a large smart factory. But what if there is no requirement for many processes, post-processes, and assembly. Everything is accomplished due to one process, the smart factory will no longer be large but small (Fig. 9.11). With the shrinking of all requirements in one process confined within one system (Figs. 9.2, 9.6, 9.9a), there are no connections required between different units or systems. Hence, there is no need for such material

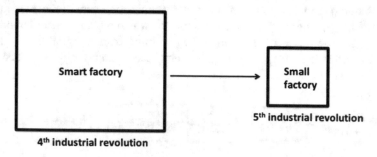

Fig. 9.11 Schematic diagram for transformation of a factory in fifth industrial revolution

handlings, floor plans, sensor based systems, big data and their analysis, or controlling and monitoring of connections between various systems. Thus there will be no issue with the failure of any connection.

A one-system factory will not make everything but will make many things [20] for which smart factory is used or will be used, and will replace many smart factories. A one-system factory is similar to one-room factory that will make few hundred parts, which is not expected to produce in mass scale. The concept for using one AM system to make a factory is not new [21].

This brings a question how these AM-propelled one-system one-room small factory will bring industrial revolution. It is argued that the role of AM will be subsidiary, its role will help achieve those manufacturing goals that could not be achieved by CM. It is again argued that since AM is advancing, its role will become more crucial, but it is not AM which is leading. Moreover, industrial revolution is about mass production and further revolution is expected to depend on factors that will increase mass production efficiently. Thus, the factors such as advances in robots, materials [22], artificial intelligence, big data, cloud manufacturing, human–robot interaction, and human factors [23] will play the role [24].

AM-led industrial revolution depends on whether mass production in many types of products is going to be outdated. If it is outdated, future revolution will be related to mass non-production. Though one-room factory is not strong and will not be able to compete with contemporary factories in the scale of production. But if one-room factory is present at every corner of the street, this one-room factory will be sufficient to cater the need of the whole street. If the street does not need many products from outside, mass production will not survive for all types of products.

Mass production for many things will stay [25] and all new advancement will augment that mass production but revolution will not be about how to compete with mass production. Revolution will be about to be free from overdependence on mass production not because AM will indirectly compete with mass production but AM will help free from the demerit of over-production. Mass production in some companies is already moving towards mass customization [26].

One-room factory at every street or locally [27] can readily be overlooked and will seem to be a novel idea requiring to be experimented. But, one-room type factory is not a new reality—this existed in the name of cottage industry or craft production before the industrial revolution, this also existed after industrial revolution because of its own merit. It does not mean it is time to find the merit of cottage industry, artisan skill and craft of the individual for ultralimited production and forget the advancement of 100 years. It only means one-room factory will fulfill the need and there is no need to look towards big factory for every need. If there is no such dependence on big factory, this situation will not be different from when every village was self-sufficient and did not have anywhere to look towards in the absence of big factory. If need is fulfilled at the lower level, it will be unfair if this fulfillment is overlooked because it gives the impression of going 500 years back.

It brings a question if AM can fulfill many of needs, what will be the role of other engineering advancements. Other advancements will play role in the mass production and big factory settings but the role will bring ongoing development in the big factory, the role will further advance how mass productions should take place. The ongoing development or further improvement in methods for mass production will not be an industrial revolution if net output will decrease in the absence of need for an increased output.

It is argued that human touch will play an increasing role in industry 5.0. Human touch [28] or increasing interaction of human with the machine [29, 30] is not a new manufacturing technology but a method to increase productivity while AM is a new manufacturing technology that is questioning the need for increased productivity in a big factory setting.

This brings a question what is the guarantee that AM will bring fifth industrial revolution, the answer is there is no guarantee. The trend in society is movement towards individualization, customization [31], personalization, sustainable less consumption, new governmental regulations for recycling, minimization of waste or waste management, awareness for environment sustainability. Seeing the trend, it is fair to observe that society is creating a vacuum where AM is fitting increasingly well. It can be a matter of guess whether AM is changing society or changes in society are creating AM. But it is not a guess that mass production of many items is getting redundant. Hence, individualization or customization needs more distributed production facilities than its absence needed.

What AM can do can be a matter of guess but what society needs is not a matter of guess. If the development of AM is delayed, an undeveloped AM will not do much help to society or industry but a developed AM will certainly contribute more. Seeing the trends in society, it is fair to predict if AM is developed, it will permeate society. This permeation is a revolution. But, why this permeation will be called industrial revolution, why not it be called societal revolution. Because the societal revolution will have fair chance to cause not to have a future industrial revolution based on mass production. But, where is the concrete evidence, there is no evidence.

References

1. Chiu, W. K., & Yu, K. M. (2008). Direct digital manufacturing of three-dimensional functionally graded material objects. *Computer-Aided Design, 40*(12), 1080–1093.
2. Gibson, I., Rosen, D. W., & Stucker, B. (2010). *Additive manufacturing technologies: Rapid prototyping to direct digital manufacturing.* New York: Springer.
3. Chen, D., Heyer, S., Ibbotson, S., et al. (2015). Direct digital manufacturing: Definition, evolution, and sustainability implications. *Journal of Cleaner Production, 107*, 615–625.
4. Holmström, J., Liotta, G., & Chaudhuri, A. (2017). Sustainability outcomes through direct digital manufacturing-based operational practices: A design theory approach. *Journal of Cleaner Production, 167*, 951–961.
5. Rojas-Nastrucci, EA., Ramirez, RA., & Weller, TM. (2018). Direct digital manufacturing of mm-wave vertical interconnects. In *IEEE 19th WAMICON, FL* (pp. 1–3).
6. Kacar, M., Wang, J., Mumcu, G., et al. (2019). Phased array antenna element with embedded cavity and MMIC using direct digital manufacturing. In *2019 IEEE International Symposium of Antennas Propagation and USNC-URSI Radio Science Meeting* (pp. 1–82).
7. Zhou, Z., Xie, S., & Chen, D. (2011). *Fundamentals of digital manufacturing science.* Cham: Springer.
8. Wu, D., Rosen, D. W., Wang, L., & Schaefer, D. (2015). Cloud-based design and manufacturing: A new paradigm in digital manufacturing and design innovation. *Computer-Aided Design, 59*, 1–14.

9. Lan, H. (2009). Web-based rapid prototyping and manufacturing systems: A review. *Computers in Industry, 60*(9), 643–656.

10. Rudolph, J. P., & Emmelmann, C. (2017). A cloud-based platform for automated order processing in additive manufacturing. *Procedia CIRP, 63*, 412–417.

11. Kumar, S. (2020). *Additive manufacturing processes*. Cham: Springer.

12. Degan, BZ., Barbosa, GF., Zubiaga, DG., et al. (2020). 3D digital manufacturing applied on drilling of 7050-T651 aluminum performed by robot. In *IMECE2020- 21914* (pp. 1–5).

13. Vock, S., Kloden, B., Kirchner, A., et al. (2019). Powders for powder bed fusion: A review. *Progress in Additive Manufacturing, 4*, 383–397.

14. Wu, B., Pan, Z., Ding, D., et al. (2018). A review of the wire arc additive manufacturing of metals: Properties, defects and quality improvement. *Journal of Manufacturing Processes, 35*, 127–139.

15. Lynn, R., Dinar, M., Huang, N., et al. (2018). Direct digital subtractive manufacturing of a functional assembly using voxel-based models. *Journal of Manufacturing Science and Engineering, 140*(2), 021006.

16. Lyly-Yrjänäinen, J., Holmström, J., Johansson, M. I., & Suomala, P. (2016). Effects of combining product-centric control and direct digital manufacturing: The case of preparing customized hose assembly kits. *Computers in Industry, 82*, 82–94.

17. Mazur, M., Brincat, P., Leary, M., et al. (2017). Numerical and experimental evaluation of a conformally cooled H13 steel injection mould manufactured with selective laser melting. *International Journal of Advanced Manufacturing Technology, 93*, 881–900.

18. Tosello, G., Charalambis, A., Kerbache, L., et al. (2019). Value chain and production cost optimization by integrating additive manufacturing in injection molding process chain. *International Journal of Advanced Manufacturing Technology, 100*, 783–795.

19. Gaub, H. (2016). Customization of mass-produced parts by combining injection molding and additive manufacturing with industry 4.0 technologies. *Reinforced Plastics, 60*(6), 401–404.

20. Berman, B. (2012). 3D printing: The new industrial revolution. *Business Horizons, 55*, 155–162.

21. Bekker, ACM., Verlinden, JC., & Galimberti, G. (2016). Challenges in assessing the sustainability of wire + arc additive manufacturing for large structures. In: *SFF Proceedings* (pp. 406–416).

22. Dilberoglu, U. M., Gharehpapagh, Y. B. U., & Dolen, M. (2017). The role of additive manufacturing in the era of industry 4.0. *Procedia Manufacturing, 11*, 545–554.

23. Reiman, A., Kaivo-oja, J., Parviainen, E., et al. (2021). Human factors and ergonomics in manufacturing in the industry 4.0 context—A scoping review. *Technology in Society, 65*, 101572.

24. Horváth, D., & Szabó, R. Z. (2019). Driving forces and barriers of industry 4.0: Do multinational and small and medium-sized companies have equal opportunities? *Technological Forecasting and Social Change, 146*, 119–132.

25. Achillas, C., Aidonis, D., Iakovou, E., et al. (2015). A methodological framework for the inclusion of modern additive manufacturing into the production portfolio of a focused factory. *Journal of Manufacturing Systems, 37*(1), 328–339.

26. Shao, X. F. (2020). What is the right production strategy for horizontally differentiated product: Standardization or mass customization? *International Journal of Production Economics, 223*, 107527.

27. Campbell, T., Williams, C., Ivanova, O., & Garrett, B. (2011). *Could 3D printing change the world? : Technologies, potential, and implications of additive manufacturing* (pp. 1–13). Washington: Atlantic Council.

28. Nahavandi, S. (2019). Industry 5.0- a human-centric solution. *Sustainability, 11*, 4371.

29. Demir, K. A., Döven, G., & Sezen, B. (2019). Industry 5.0 and human-robot co-working. *Procedia Computer Science, 158*, 688–695.

30. George, S. A., & George, A. S. H. (2020). Industrial revolution 5.0: The transformation of the modern manufacturing process to enable machine and man to work hand in hand. *Journal of Seybold Report, 15*(9), 214–234.

31. Yao, X., & Lin, Y. (2016). Emerging manufacturing paradigm shifts for the incoming industrial revolution. *International Journal of Advanced Manufacturing Technology, 85*, 1665–1676.

Index

Printed in the United States
by Baker & Taylor Publisher Services